UL

SELLING

The Art and Science of Sales Success

Jim Martin

How simple changes can
turn everything into sales success

Published by:
Ultimate Selling Solutions, LLC
P.O. Box 405
Stevenson, MD 21153-0405

Tel: 410-878-2663 Fax: 888-325-4001
Website: www.ultimatesellingsolutions.com

Ordering Information:

Individual sales: Available through most bookstores. They can also be ordered direct from the publisher/author online @ www.ultimatesellingsolutions.com

Quantity sales: Special discounts are available on quantity purchases by corporations, associations, and others. For details, submit your request to the publisher/author at the above address, or submit email to: info@ultimatesellingsolutions.com, Subject: Volume Book Purchase.

Library of Congress Control Number: 2011923613

Martin, Jim, 2011

Ultimate Selling: The Art and Science of Sales Success / by Jim Martin

ISBN: 978-0-9833054-0-8

Printed in the Unites States of America

Foreword

In my long years of working, I've met a lot of people who sell or manage sales people. Many of those people thought of themselves as experts. Jim Martin is the only one I have ever known who has done it and managed it in a number of diverse organizations, successfully strategized about it for others, lived and breathed it. Now he has written a book that documents his process for the rest of us to use for our success.

This book is different. It removes much of the "gee whiz' about selling and shows that the practice of sales is like everything else that works—a discipline supported by hard work, preparation, patience and integrity. The concepts are simple. The work to employ those concepts is not easy. If you are willing to do the work, you will make money, save time and build customer relationships that should last as long as you are working. And there is the key: selling with effort, selling with a customer-needs focus and selling with integrity builds relationships that can last a long time with great mutual benefits to the parties in them.

This is just not a book you read once and set aside. It is a reference book that you should consult frequently. Use the materials in the book to create a sales success infrastructure that wins the success you desire for yourself, your colleagues and your company. Learn, have fun and do the good work that keeps our communities and our society going.

Paul A. Riecks
Running Your Own Company: An Entrepreneur's Field Manual

WHO SHOULD USE THIS BOOK

While many of the concepts and ideas in this book may appear familiar. I trust you will find their application to be significantly different. Our method is actually the *opposite* of most selling methodologies in use today.

Ultimate Selling focuses on the needs of salesmen and women who want to be true, productive selling professionals. It's also a useful tool for sales managers and teams. Many of the concepts apply directly to C-level executives, since achieving a high success rate for new business development are top down objectives.

Sustainable growth and profits are the desired outcome. Long-term customer retention is assured when a sale serves your customers' vested interests, today and beyond. We teach selling with both immediate and residual value in mind.

Increased profitability is also a direct result of reducing expenses and resources in pursuit of opportunities with little or no chance of success.

With the change in buying options and challenges that the internet presents to selling organizations, we believe that a contemporary process must include selling techniques designed to attract customers on factors beyond price.

Ultimate Selling works. It will have positive effects on individuals and departments across an organization. How well it works will be determined by you. If you apply the tenets of Ultimate Selling, I'm extremely confident you or your organization will never go back to your current way of pursuing sales. Ultimate Selling is all about maximizing successful outcomes! It's about the pursuit of perfection. Nothing in sales is perfect but we'd rather pursue perfection and achieve a higher level of success.

This book provides you with specific guidance about what to do and say when involved in selling. It's definitely not theory or fiction, but the result of years of experience in real world selling.

Becoming an Ultimate Seller is like a large puzzle and as you discover how each piece fits, you will get closer to discerning the whole picture.

Best wishes for great reading, great application and the ultimate outcome: great selling.

—Jim Martin

Table of Contents

Appendixes:

Acknowledgements

A special thank you to family and friends for all their support and to my business associates and clients, who over the years have encouraged me to put my knowledge and methods into print. I deeply appreciate all the salespeople and organizations I've been associated with over the years. They were a tremendous source of firsthand knowledge, experience and the genesis of Ultimate Selling.

To my copy editor and associate writer Katie Ritter, for her hard work and many contributions. Katie's exceptional writing skills helped me condense many years of successful selling experiences into a concise book. I appreciate her understanding of my vision for this project as well as her unflagging love of grammatical detail.

Thanks to the professionals at Absolute Services: Marty Ostendorf, Teresa Exley, Danny Piano, Lyndon Famularcano, Harold Medina, Neil Tanguilan, and April Pascua. Without their expertise and keeping me on track, this book would not have become a reality. I'm forever grateful.

Adam Noll, for his hard work and contributions to the design and branding of this book. Adam has helped to beautifully reinforce the design and layout of Ultimate Selling. I appreciate Adam's ability to translate my experiences into a easy to follow visual presentation.

To Paul J. Meyer, Founder of Success Motivation Institute (SMI) for creating a business opportunity where David H. Sandler and I were able to contribute and grow professionally and personally. In memoriam, I would like to pay tribute to Jim Sirbasku, President of Profiles International and former vice president and president of SMI. Jim's ongoing support as well as those of the entire SMI network of trainers and motivators was inspirational.

Finally, I'd like to extend my sincere appreciation to the executives and seasoned sales professionals from diverse industries for their contributions as an editorial review team: David Bellos, Josh Farace, Mike Gill, Jay Litton, Gil Cook, Bruce Manthey, Phil Monetti, Marty Ostendorf, Paul Riecks, Mark Ritter, and David Webb.

Dedication

To my brother John, In memorium

To my daughters Melissa and Jennifer

To my wife Edna

Chapter 1

What Got *You* Started in Sales?

We all have a story. Here's mine.

A sales career requires mastery of certain skills to be successful—or you fail. I began my selling career many years ago as a kid determined to make money. I was the youngest of eight children. If I wanted something, there was plenty of competition for it.

One family story relates that I would get up in the morning, make egg sandwiches, and then go to the neighbors to sell them. That progressed into selling greeting cards. During holidays, I'd sell flowers or Christmas trees. Ambitiously, one year, I cut down an evergreen tree in the neighborhood and went door to door with it until I pocketed some money. (I think I've been forgiven by the owner!)

Fast forward to my first real sales job as an account representative with Campbell's Soup. I called on local, regional and national grocery stores and soon discovered the job was as much about product management as it was sales–but I can't overstate the value of the experience.

It's fun but odd to look back, because Campbell's utilized sales representatives in their just-beginning concept of brand-recognition strategy. While sales was an objective, it was secondary to addressing product placement and visibility.

A typical day would include visiting fifteen or twenty stores and checking on all products, from soups to beans to juices. If out, we personally pulled inventory from the back room, priced the product and re-stocked the shelves. Making sure we had proper positions of products was a key task, as was ensuring full product-line representation. After we finished, we met with the grocery or store manager to get orders for out-of-stocks, sell displays and present incentive programs.

Daily work requirements, high expectations and personal accountability became important foundations of my sales job. Campbell's audited our work carefully. A corporate representative would arrive unannounced in our territory with our sales report and actually replicate the day's activity to audit the work. I had to be extremely focused on my responsibilities, and attention to detail was a very important component.

Seque: in a grocery store one day I met David Sandler. David had been in the food distribution business and now was a local representative of the Success Motivation Institute. We struck up a conversation, went for coffee, and several meetings followed. Eventually I joined his business on a part-time basis. It wasn't too long after that I left all the security of Campbell's to join David and Success Motivation. Don't ask me what I was thinking at the time: I had a wife, one daughter, another on the way and in the move I would lose my salary, health care, company car—and all of my job security. That was a leap of faith!

A Challenging Beginning

Was it ironic—or moronic?—to be selling success and motivation, given that I had just started in the 'sales' business? I remember an early experience in which I was parking my financed Volkswagen for a scheduled appointment. Next to me was a sharply-dressed businessman in a luxury automobile. We exchanged morning greetings and rode in the same elevator. You guessed it: he was my appointment and the owner of the business, to whom I would be providing 'valuable' advice.

My daily schedule with Success Motivation required me to make fifty dials and thirty cold-call walk-ins. Activities such as daily and weekly planning, appointment preparation, acquiring product knowledge, and other non-sales activities had to take place after hours, as I could not afford to waste prime 'selling' time. My pay was 100% commission on ***money received***. There was no draw against commission, no base salary, no bonuses...and I even had to pay for all my own business expenses and benefits!

I really looked forward to days on which I had scheduled appointments, because I'd then only have to do twenty-five dials and fifteen walk-ins...but unless I had a minimum of five face-to-face appointments, the twenty-five calls and fifteen walk-in daily requirements were inflexible.

Here's the thinking, with which you may identify: how else could I build and maintain enough prospects in my pipeline to ensure enough sales? My daily activity quotas were based on the "ratio thinking" of conventional selling, rather than focusing on key activities that would produce better outcomes.

It's as grueling now to be faced with those kinds of requirements as it was then, even with the assistance of technology tools.

Was I just stupid to make this leap—or just ignorant of the impact? Fortunately, it turned out to be the latter. After all, as the saying goes, stupid is forever. Ignorance can be fixed.

Had it not been for a strong daily work ethic, desire to survive, and blissful ignorance, it's doubtful that I would have made it. I struggled. There were days when I needed a cash deposit to pay for parking. It reminds me of my days in the Marine Corps Basic Training at Parris Island in South Carolina. I would not trade the experience for anything—but I'm sure not interested in going back! It's important to note that like boot camp—and like so many other experiences in life, I did not have to like it, I just had to do it.

Fortunately, I was able to get in the game and be at the plate enough time to figure out—by trial and error—how to succeed. And while they can be very painful at times, it's the failures that teach us the most.

My years with David at SMI were a life-changing experience. We were fortunate to share many successes—and yes, some failures—as we worked on always pursuing a better way. Our contributions to the Success Motivation Institute were significant enough that we were the recipient of repeated awards. In fact, we got so good that Paul Meyer, founder and president of SMI, sent his son to Baltimore, so I could take him on the street and demonstrate our systems for cold calling and selling.

Success in my early days was about the will to win: the tenacity to simply hang in long enough until I had acquired and mastered the skills to succeed. It was about the ability to survive financially while learning the craft of sales.

My career progressed. As I gained experience, I found myself always challenging the existing paradigms and asking myself how can we do it *better*?

I realized that certain elements had come to be the cornerstone of my personal sales success formula. Over time, it made sense to formalize my techniques so that they could be shared with others. Approaching sales in such a way that you can dramatically improve your sales is something I came to call "Ultimate Selling".

This book is my contribution to the selling profession. I've been in your shoes. Your time is valuable, as are your company's resources. As you explore Ultimate Selling, one of the most important skills you'll learn is how to turn selling dynamics around so that *prospects must earn your investment of time and resources rather than you pursuing them...and in addition, they will want to work more and more with you and your company.* My goal is to help you avoid some of the failures I endured, and to accelerate your career to new levels by sharing my experience. I want to give you the tools of Ultimate Selling in this book.

To do this, we'll show you a new approach to qualifying opportunities, to selling 'up front', and to becoming a trusted advisor to your clients and customers. We'll give you our best practices from start to finish. We hope not only that you will learn the best way to do things, but also that you will actually *enjoy* Ultimate Selling.

What Ultimate Selling ISN'T

Before you discover what Ultimate Selling is all about, here are some insights on what it's NOT about:

- It's not about prospecting
- It's not about social networking
- It's not about "getting past the gatekeeper"
- It's not about lead sourcing
- It's not about becoming best friends with a prospect
- It's not about sales force automation or CRM programs or any of the plethora of technology tools
- It's not about seminars or webinars
- It's not about "17 Super Closing Techniques!"
- It's not about traditional sales pipelines and forecasts
- It's not about giving free education to suspects or information-gatherers

While many of these subjects support selling, they're just tools. They're not a prospect management strategy.

You can have all the tools in the world, but without a construction plan-a strategy-you can't build a house.

Ultimate Selling is primarily focused on accomplishing specific objectives from very high-quality interactions, whether over the telephone or face-to-face. It's not about technology. It's not about tips. It's not about tools. It's about you engaging your prospect using a meaningful, deliberate strategy.

My goal is to have you master Ultimate Selling. While you will probably use things like social networking or daily sales plans, Ultimate Selling requires that they are simply used as tools in the Ultimate Selling process.

Many of you may also have in-house systems with their own selling rules. At best, they may be tired and marginally effective. The strategy of Ultimate Selling will enable you to move beyond what they offer while working within them.

Be inspired to work at and master Ultimate Selling! If you do, I promise you will *never* revert back to traditional selling practices.

The Evolution Of Sales Methodologies, or How People Have Sold

To appreciate the value of Ultimate Selling, and before you can understand our expectations of you, let's spend a few minutes reviewing how sales methodologies evolved over the years. *Where are you in this evolution?*

Push-Through Sales and Win/Loss Ratios

Historically, conventional selling systems were built on "push-through" methods of selling. The salesperson did all the talking and presenting, and when the prospect objected, they would proceed with what was termed the "Stalls and Objections" procedure, whose steps were:

1. Listen carefully to the prospect's objection
2. Convert their objection to a question
3. Answer the question
4. Ask a closing question
5. If they object, begin again at Step 1

This repetitive process was intended to overcome their objections and remove any buyer hesitations, all of which were expected to come up right at the end of the process. The belief was that you if you handled them and went for the close…you had a sale. You'd won, you gotten your check, and you didn't much care what happened after that. If not, you'd loop through the process again…and again…and again.

This sales methodology was built on win and loss ratios, typically with far more losses than wins. It was, as you have no doubt heard at one time or another, a 'numbers game'. The goal was to make ten presentations and close two of them. If you were really good, you might close as many as three or four. Using these numbers your closing rate would be twenty percent; if you had a great month, thirty percent.

But that meant you were planning to *lose* sixty, seventy, eighty percent or more of your opportunities! The genesis of Ultimate Selling was our rejection of the idea that planning to lose that often was completely unacceptable. There had to be better way.

Imagine interviewing as a salesperson for a job and hearing honestly about traditional selling statistics. If someone suggested to you that you would get to lose 70 or 80% of the time in order to win only 20 or 30%, how excited would you be to join the selling profession?

Sometimes in the push-through model you would accidentally develop a long-term customer if your product or service was exactly what the customer needed. This was more accident than plan, because push-through selling wasn't about the customer or client: it was about the sale and getting the money!

That's not to say money isn't important, because it most definitely is. But you can develop both the relationship *and* make it to the money each time when you apply the tenets of Ultimate Selling.

Next Stop: Consultative Sales, Stage One

Fortunately, the sales profession progressed to a newer sales model called 'consultative selling'. There were two evolutions of consultative selling. In the first stage, consultative work was still 'seller focused'. Salespeople attempted to understand what the prospective buyer wanted and needed—but as with all change, it was difficult for many to master. They would revert back to push-through selling as it was more familiar and comfortable. As a result, it was still more about the salesperson's goals, and customer expectations weren't met very well.

Next Stage of Evolution: Customer-Centric Selling

This evolution took consultative selling to the next natural level. It is a truer model as it is more customer-focused: understanding the buyer's needs, and having your product or service fulfill that need.

Customer-centric selling focuses on achieving a buyer's goals, solving their problems and satisfying their needs. This sales model incorporates a high level of integrity. Your experience and your depth of knowledge about products, the market and business acumen play an integral part.

While customer-centric selling is a significantly improved selling model, it requires a highly skilled salesperson. Doing it with good intention but little skill puts much of the control in the prospect's hands, and puts many salespeople at a disadvantage.

Welcome to the Twist of the Internet Age: *Buyer Advantage, Seller Beware*

You can be assured these days that your prospective clients have done price checking on the internet and are savvy about ways to reduce their price and your profit. Websites make all kinds of pricing and product reviews instantly available.

Welcome to the new game, where selling seems to be all about the lowest price. At the same time, the cost of selling—from advertising to marketing to acquiring prospective customers to working with them until a sale happens—is enormous, and increasing. Consider the costs that are often squandered in the pursuit of unqualified prospects!

Remember the adage of seller advantage, buyer beware? In today's information-accessible era, the new paradigm is *buyer* advantage, *seller* beware!

Given this competitive marketplace we recommend that you invest your sales resources in opportunities that are productive: those *having defined imminent needs*, with *defined timelines to buy*, and *a defined way to pay*. Getting your prospects to tell you their imminent needs, their real timeline to buy, and their way to pay is the art of Ultimate Selling. Lacking those criteria, you will see your cost of sales go up and your loss rate accelerate. With it, you will see weak prospects fall aside quickly, and strong prospects developing.

Besides *How* We Sell... What Are The Other Problems?

In all my years with selling, businesses both large and small are always challenged in four ways.

First, they are far too reactive in the sales process. A prospect starts looking, and the business jumps to convince the prospect they have the right solution. Time is not taken to vet the opportunity...which puts the prospect in control of the sales opportunity, not you.

Qualifying needs to be a fearless process. We're going to give you courage to do it ridiculously well...with the expectation that you will then <u>only</u> work with prospects who are <u>truly</u> qualified and deserving of your time.

Last and most important, businesses have accepted too many failures in sales pursuits as an acceptable part of their selling process. It's almost as if there's a status quo of "lots of failure is how we do it here". This is reminiscent of the famous quote about insanity: you can't keep doing something the same way, and expect different results.

For many, if not most, sales reality is that numbers are the solution: just have a big pipeline and issue lots of proposals. While numbers are important, if you apply significant, true qualifying to the formula, your ability to acquire new business will improve. Likewise, use of company resources will improve per sale which can have a positive impact across an entire organization.

Imagine what could happen if we first understand the prospect's goals, issues, challenges, problems and preferred solutions, but not in a cursory way... at a much deeper and more powerful level?

What could happen if we developed simple ways of *turning around each interaction* so that *we* were in control of the decision instead of them?

What would happen if by the time we presented our proposal, it was a mere formality, and all decision-makers had already indicated their intention of accepting?

What if every interaction with prospects was a way of gaining knowledge, of being more and more helpful to them, continuing to build a strong foundation of trust?

We would have become very proactive in the art of Ultimate Selling instead of reactive, and we would have crafted the most expedient, efficient way of winning their business.

My desire has always been to pursue perfection and be 100% successful in my sales outcomes. I considered anything less to be a very poor return on both my investment of time and the use of my company's resources. In pursuing a 100% success rate, I had to keep refining the steps to the ideal sale. Here's what happened: I ended up turning the whole push-through, stalls and objections process in reverse, kept a customer-centric approach, applied the art of honest communication with the discipline-science-of a new strategy, and Ultimate Selling was born.

Problem Two: there's never a 'right' time to improve. Many times I've observed organizations and individuals reaching out for help when they were in dire straits, somehow hoping for a miracle. These were usually the very same ones who didn't need anything or even want to consider investing in change improvement when business was great. When things were good, they were just too busy or had an "if it isn't broken, don't fix it" mentality.

But the time when things are good is the best time to invest in improvement, both from an emotional and financial standpoint. Almost anyone can succeed when the market is hot and the economy prospering. Smart planners always work at doing better, and don't wait until the demand is critical.

Problem Three: little deceptions send the wrong message. Let's be honest with each other.

I constantly find it interesting that salespeople and organizations are always apologizing for selling. They're not open enough or appreciative of their sales staff's true and valued role in new business development, either to their clients and prospects or to the sales staff themselves.

How often do you see a business card that indicates the bearer does professional sales? It's always 'Account Manager' or 'Territory Representative', or something that makes sales look like something to be tidied up and covered over.

I've never hidden or apologized for the fact that I'm in sales. Instead, I made it very clear to my clients and prospects that I would work very hard on their behalf. I was willing to earn their business. If our product or service wasn't something that provided value to them, I expected a no-sale outcome. That's simply being professional.

That may be a tough pill for a salesperson to swallow, but being able to say 'no' when it makes sense to do so is the only way you can arrive at a *true* win for both you and your prospect while earning their trust. This should be the case regardless if you have full pipeline of qualified prospects (it's a little easier then!) or if you have none. You have to do the job right.

By always being honest, you never have to try and remember what you did or did not say. It's another way of ensuring that character, integrity and ethics are embodied in the sales relationship. Your honesty makes it possible, even comfortable for your prospects to be honest with you. It also gives you the right to expect honesty and integrity from your prospects, and to say so to them.

Enough history. Let's get started finding out how Ultimate Selling works.

Chapter 2

What Is Ultimate Selling? Setting Your Expectations

At the core of Ultimate Selling is true customer-focused, consultative selling that embraces essential ingredients of all successful selling methods, but adds the Ultimate Seller's rules of engagement. So what makes it so different—and valuable to you?

1. ***Ultimate Selling changes the order in which things happen.*** In most sales, there's a lot of energy spent getting the proposal created and a great amount of tension getting it accepted. In Ultimate Selling, there's almost *no* energy in the proposal. Instead, a lot more happens up front. Closing a sales becomes almost an afterthought…a given.
2. ***Ultimate Selling puts you in charge of the process.*** A lot more happens in vetting prospects, more than you've ever done before. You choose them instead of them choosing you. Sound a little unconventional, It is, and it works.
3. ***Ultimate Selling requires clarity and honesty,*** both from you and your customer, as an integral and essential part of every interaction. Agreements are made and honored. The "trusted advisor" term is used frequently in sales books. We're going to teach you step by step how to *actually become* one. Since your goal is to genuinely be a trusted advisor, this entire approach is wrapped in a blanket of ethics and integrity.

That last item is the 'art' of Ultimate Selling: honest communication. Art is just skilled workmanship. Do you regularly talk to your prospects honestly, in a skilled, comfortable way? We are going to teach you how to do that—and it can change everything.

Let's talk about the 'science' part for a moment. Science does <u>not</u> mean 'complicated' or 'hard or 'technical'. Science, at its basic level, is just *knowledge gained from observing and experimenting.*

Building a house can be an art and a science—as a set of skills that have been tested and perfected over time. Improving athletic performance can be art and science: a platform diver observes, experiments, works at skills—and executes a double back flip with a twist.

So the science part of Ultimate Selling is *knowledge you get from observing, learning and experimenting.* Couple that with *applying skills you've honed.* You work with prospects, you pay attention, you learn. You experiment with what works, you grow in skills, you apply those skills. That's your personal art and science of Ultimate Selling.

Like any good science, an acronym is helpful and fun. In Ultimate Selling: P + Q + N + C = C. What does that mean? (No, there will not be a test.)

Ultimate Selling is a very simple process that is tightly focused on specifically what to do when you are interfacing with prospective buyers. So how does this happen? Let's take a visual tour of some of the chapters to come. In the following table there are simple elements that we all know help develop relationships with prospects (and win sales).

Look down the left column. You'll see the skills that form the foundation of our science. They are simple ones: Preparation. Qualifying. Negotiating. Confirming. Connecting.

That's the structure upon which all of Ultimate Selling is based. Preparation + Qualifying + Negotiating + Confirming = Connecting. Simple stuff, used differently, can become very powerful: you can use a shovel to dig a hole, or you can use it as a lever to move huge objects. Both valid. One more sophisticated.

Read this chart carefully....You'll immediately start to see a little of how Ultimate Selling turns things around.

Table 2-1 The Ultimate Selling Process

Arena	How We Do It Differently and Why It's Valuable
PREPARE	You're ready to handle anything that comes up as you move through the process. Think about being prepared. Good preparation allows you to be calm. Everything is anticipated and available. Being calm means you're more likely to control the situation. You're listening and asking the best questions. Whether your typical sale happens very quickly, or over a period of several months, being prepared and ready keeps you feeling competent, and it reflects that way to your prospects.
QUALIFY	We'll help you discover early in the process about the true probability of a sale, rather than waiting until you submit a proposal to find out if it's yes or no. Why so early? Well, what's the point of doing the work and submitting a proposal that may be doomed from the beginning? You won't qualify prospects: you'll *QUALIFY* them! You'll learn to ask boldly up front when they will take delivery or deploy your service or solutions, how they will pay for it, and how the buying decision will be made.
NEGOTIATE	Negotiating begins as soon as you uncover issues or challenges that your prospects are facing. It may be with your products or services, or it may be internal problems in their business. Whatever the issues, you will learn to work together with your client to create mutually acceptable terms and conditions of the sale. This will happen verbally. They will design the compromise, and it will happen well *before* you present your value confirmation.
CONFIRM	You will demonstrate, in *their* language, using words *they have supplied for you*, how all of their concerns, objections, obstacles, needs and desires have been addressed. Confirmation is a relatively minor process which replaces traditional proposal presentations. You'll learn later how to do all the hard work up-front, long before this step.
CONNECT	You'll connect with this now-client in a meaningful way, creating the foundation for a long-term relationship as their ultimate 'trusted advisor'.

Now lets look at how doing those skills as part of a disciplined strategy strengthens your relationship with your prospect. The first two columns are what you do. The third column shows your changing status *from the prospect's perspective*. Will teach you how to use those skills to connect you tightly with prospects.

Table 2-2 Competent to Trusted Advisor

Knowledge And Skills	Acquired Via	How They See You
Know your company's strengths and challenges	Preparation	Competent
Know your product or service Know your competitor(s) products/services, and how you compare to each other	Preparation	Competent
Know what it takes to close your typical sale	Preparation	Competent
Learn your prospect's business and who's who	Preparation	Professional
Know how to develop a flexible game plan for every step	Preparation	Reassuring
Learn your prospects's goals and needs by listening and asking questions	Qualify: Interest	Advisor
Discuss your prospect's ability to pay	Qualify: Money	Advisor
Learn your prospect's decision making process	Qualify: Decision	Advisor
Find out your prospect's timeline to buy	Qualify: Decision	Advisor
Identify and resolve areas in which what you offer does not match the needs and and wants of your prospect.	Negotiate	Trusted Advisor
Confirm to your prospect your understanding of their needs and wants, as well as any issues you had to negotiate	Confirm	Trusted Advisor
Begin the process of them getting what they need and want	Connect	Trusted **Provider** (they are now your client)

Now, let's address any doubts right away. This chart may look all well and good, but you may be dubious if Ultimate Selling really works as well as we say it does. Or it may look like too much preparation work once you get to those later chapters that isn't really of any value.

Time for a True Story

I had a lucrative sales opportunity, but with obstacles many would think impossible to overcome. As I share this story, think of how you would have rated the opportunity. Would you have believed the obstacles could be overcome? *Would you have pursued the opportunity?*

A large health-care organization had a vendor who had been supplying services to them for many years. It was such a long relationship that the vendor actually had two vice-presidents of their company serving on the prospect's board of directors! Additionally, an advisory board member owned a similar business. With two competitors so well entrenched, would you think we had a chance at the opportunity?

Most people would think that if the organization was unhappy with the very-secure incumbent vendor, the obvious next choice would be the company owned by the advisory board member. Well, that's not what happened.

The existing vendor was replaced by *our* company and their two board members chose to resign. The advisory board member pulled out all the stops to get the sale for his company, unsuccessfully. The initial sale was $1.3 million at a cost of $650K. It also generated several years of recurring revenue that exceeded a million dollars. For those who like to count dollars, the commission portion was 5% of gross revenue, or $65,000 dollars. We were able to unseat not one but *two* favored vendors by applying the simple, effective principles of Ultimate Selling—to our profit.

Are you intrigued about becoming an Ultimate Seller? Then let's proceed. First, let's start on familiar ground. We use familiar skills in powerfully—different ways. What do they mean in Ultimate Selling language? Why does it work and how?

First, Ultimate Selling Works Because It's Efficient

The more you can address up front, the easier it becomes on the back end to eliminate waste and re-work. You will avoid the trap of squandering time, money and resources that should never have been allocated.

Let's look at an example of how the re-engineering process was applied to auto manufacturing by Lexus to successfully launch a new, high end luxury division and compete with Mercedes-Benz.

The Lexus Advantage: *Doing The Work Up Front*

When Lexus was first introduced, people kept asking the question: how could Lexus build such a high-quality car—that sold for so much less than a Mercedes? What was their secret?

While Mercedes-Benz is a high quality automobile, they touted rigid inspection procedures and quality control… and at the time, much of that was on the ***back end*** of the manufacturing process.

In fact, their early brochures illustrated how the vehicle would be repainted until it was right. In manufacturing, this is called "rework", and while it can produce a desired quality outcome, the process of re-work is very expensive for both labor and material costs.

> ***Pause for Thought:***
>
> ***Can you think of any examples of re-work in your pursuit of sales? Think of when you've had to create a second, third or fourth revision of a proposal with revised pricing each time options are changed.***
>
> ***Wouldn't it have been more efficient in the long run—and less stressful—to have uncovered and resolved these issues up front?***

Lexus re-engineered the manufacturing process and established stricter tolerance standards with suppliers. They looked for places where costly errors could occur and engineered them out of the manufacturing process. Quality control was designed engineered up front into the build process, instead of correcting errors expensively on the back end.

At the end of the day, paying higher prices for quality components and re-engineering their process to build it right the first time resulted in many gains: higher quality, increased production, higher margins and customer satisfaction, loyalty, and long term growth. Think about how Lexus quickly evolved into a premier luxury auto provider.

The branchild of the Lexus process was Dr. W. Edwards Deming, now known as a quality guru—who had been previously rejected by U.S. auto manufacturers.

Today, Deming's principles support the global success of Proctor & Gamble, Ritz-Carlton and many more. Deming's profound yet simple strategies offer organizations a proven system to achieve lasting growth and prosperity. The principles apply universally to any enterprise—and, in fact, to you and me personally.

Think how great it would be to have your sales process work the same way. A smoothly-working process leading to a consistent quality result sounds like something we all want from our sales efforts, doesn't it?

Let's change how you work to be effective and produce better results with less backend work—just like Lexus reworked the manufacturing process.

You're going to get better at working up front, identifying quality control issues early in the process. and eliminating them as early as possible. That way they don't surface late in the game to undo the sale or make it difficult to win.

Lexus demanded quality components to support their efficient manufacturing process and accepted nothing less. Like Lexus, Ultimate Selling also works because it's about well-prepared, solid, components that fit smoothly into a well-designed process.

Ultimate Selling **teaches you to let prospects create the solution they want with you.** When it's exactly what they said they wanted, when they know they created it, they are strongly attached to it—even though it was created through your direction and carefully controlled process.

Want to know a little secret? Your prospect will totally embrace your more deliberate, careful sales process…because they will come to understand that it's in their best interest, not yours.

We hear you thinking *"This makes sense. But what's it going to take to get me from where I am….to being an Ultimate Seller?"*

My goal is for you to *own* Ultimate Selling. I want you to absorb it so thoroughly (with practice) that you are at your peak performance applying the tenets of Ultimate Selling with every prospect and every interaction.

I understand that you may be tired of working so hard for frustrating results. And I'm confident that Ultimate Selling will give you real gratification: clients who appreciate your efforts, good financial return, the respect of co-workers, and enjoyment of your work.

If you were confident that by converting to Ultimate Selling processes you could win 100% of your sales initiatives, you might be ready to jump right in and roll up your sleeves. However, you haven't used Ultimate Selling and you haven't had it proven to you, so our challenge is to change the way you look at sales-and that will take work on both your part and ours.

Whoa! Change is threatening and difficult and emotional. Why might it be hard? Because as salespeople, we want instant and immediate gratification. If you think about six-pack abs, you don't get great abs in thirty minutes. You get them from doing thirty minutes every day. It's not complicated… but it is work.

Getting Ultimate Selling right will involve some re-working of your sales process. How much will depend on the gap between how you sell now compared to Ultimate Selling strategies. This could be the hardest part of adopting Ultimate Selling, but I'm committed to working through it with you.

Is a 100% close ratio on your sales efforts achievable for you? I believe it is. But if you don't, would you accept 80%? I suspect even a 70% success rate en-route to a 100% one is much higher than you're achieving today! I ask you for two things: willingness to suspend any disbelief that it won't work…and willingness to really work on this. Begin by creating your own self-fulfilling prophecy that you can do it.

"The only thing that can guarantee success of a doubtful undertaking is the belief in the beginning that you can."

-attributed by Paul Meyer to William James

Let's get started with going over quickly what your personal foundation looks like, just to see a little about how it eventually will support your sales efforts. In addition to the chapter text, you'll find some brief practice assignments to be completed in the next section.

Be sure that you complete the assignments to the best of your ability. And, don't hesitate to reach out to others who may provide information and insight. It's about you, right now. Don't short-change yourself: do it right!

I've been in your shoes. I believe you can do this. I'm right here with you. Game on.

Chapter 3

Preparation: Laying the Foundation

Why Prepare?

Preparation is critical. It sometimes turns out to be a matter of life or death. In sales, it's most definitely a matter of quality of life for you and your loved ones.

We showed with the chart in the last chapter that good preparation allows you to be calm. Being calm means you're more likely to control the situation. Controlling the situation means you're ready to handle anything that comes up as you move through the process.

The value of preparation holds true, regardless of the situation. Let's step away from a sales perspective for a moment and reflect on how important thorough preparation counted in the following stories.

Recall the incident in 2009 in which Captain Sullenberger on Flight 1549 landed a US Air Airbus A320 in the Hudson River with zero fatalities. As a private pilot, I can certainly identify with emergency procedures. When all power is lost, you no longer create lift and you're coming down fast! There is almost no time to think about what to do.

Two significant factors played into the final outcome of that flight. Depending on your perspective, the fact that they had water beneath them instead of city buildings and no traffic in the water constituted either divine intervention or incredible luck.

The second factor was the steadiness of Captain Sullenberger and his crew in that instant. Their constant, long-term preparation created an opportunity to use the Hudson's open water below as a landing site. All pilots constantly go through checklists and emergency procedures. This event happened in a split second and required immediate responses. Preparation fed the automatic and instinctive responses to put the plane down safely.

With all the sophisticated engineering, planning, precision, and back-up systems, things can still go wrong. It's not often, but when they do, hopefully you can prepare to be your own "Captain Sullenberger" at your sales events and interactions.

Ultimate Selling is about the goal of succeeding 100% of the time, not 10–30% of the time. That takes practice so that it's automatic and instinctive. But doing professional sales isn't a life or death situation…or is it?

Success or failure on a sales call is not life or death. However, if you think about your financial well-being, the well-being of your family, the ability to educate your kids at preferred institutions, acquiring good healthcare, having people employed, being an employer, providing for necessities or for your favorite charities, it's definitely about living life well…or not so well.

Selling is like competitive sports and the best in each prepare constantly. Let me give you an example to inspire you. More than likely you've heard of Peyton Manning, premier quarterback of the Indianapolis Colts. Jim Mora, who coached both the New Orleans Saints and the Indianapolis Colts, was quoted in a newspaper article about what he believes sets Peyton apart, and in a word, said "It's preparation". Manning prepares mentally, physically and emotionally to be the best he can be, knowing that doing so allows him to be more in control of the game.

Now, everybody wants to be the best—but Manning has the drive to actually do what it takes. As a result, it's extremely rare that you see him in anything but complete control of the situation.

Switch to a sales scenario: too many times you see unprepared salespeople who are caught off guard revert to a "tap dance" response, scrambling to find quick ad-hoc answers, saying anything under the pressure of the situation. Does it sound uncomfortably familiar to you? It's not a winning scenario. Scrambling for answers, realizing suddenly that you've blown your chance to reach the decision-maker, or that they really didn't have a budget… it's not fun. Or profitable. No matter how well you tap-dance, your client can sense that something is awry.

It also means that your prospect is calling the shots and you're just reacting, hoping to jump over the next hurdle, hoping you'll get lucky and win a deal.

That would never be Peyton Manning. His extensive and constant preparation has been done often enough that it is automatic and instinctive. Your ability to be very well prepared, with objectives you intend to accomplish, with a strong and honest attitude, and with needed facts and information, and ready to engage with your prospect at any time on any point in the sales process is critical. Anything short of being fully prepared can quickly dilute your opportunity for a successful outcome.

The question is not *can* you decide to prepare like Peyton; the question is will you do what it takes to be the best? No doubt you've heard "plan your work and work your plan". Ultimate Selling takes sales planning to an entirely new level.

Sounds like work. What can you give me right now that I can use?

Being prepared enough to be calm will make you appear strong and competent. That's attractive to prospects.

Look at it this way: when you're not prepared, you're nervous, always worrying what could go wrong or what might come up that you're not ready for.

When you've prepared thoroughly, you can relax. Instead of worrying about yourself and any possible failure or embarrassment, you can focus totally on your prospects. You can let your knowledge and professional selling skills flow naturally and truly listen to the discussion at hand.

When you incorporate integrity as well, you will not be thinking about what you need to say next or how you will respond to their statements or questions. It will free you to really care about their needs. It will free you to rely on and trust your knowledge and skills. The ability to focus on the prospect and listen intently will allow you to quickly grasp and understand what they value…not what they say. It will be your trigger to ask confidently for more clarity when required.

Being relaxed will help your prospect to be relaxed. They won't feel the pressure typically associated in the buyer/seller scenario. If they're relaxed and not feeling threatened, they will be more open in their communication with you—and that's the frame of mind you want them in.

With your mind calm and clear, you're ready to use every interaction with your client as an opportunity to deeply grasp more about what your prospect wants. This allows *you* to control the situation and respond to their needs appropriately.

Controlling the situation, be it a football game or sales, is key to a successful outcome. Ultimate Selling is going to teach you how to seriously improve your control of the situation.

Think back to your last meetings with prospects or clients—were you TRULY listening to what they were saying? More importantly, WHY they were saying it? What did it mean?

Why weren't you? Chances are you were not fully paying attention because you were thinking ad-hoc about how to react to whatever was going on. You weren't being Peyton Manning, planning a strategy and practicing tactics long before the game. You were being the other quarterback, reacting defensively.

Which quarterback do you want to be?

Yes. We thought so.

So how do you get there? We don't run drills or throw footballs or review game tapes in Ultimate Selling. But we do have specific skills for you to learn, and areas of preparation that will give you a stronger foundation, a kind of core strength to support all your sales efforts.

We're not, however, going to get started on teaching you preparation skills right away. These skills take discipline (like Manning's training does!) and will be taught in later chapters. We believe you will be better served by learning the basic concepts of Ultimate Selling first. Then, when you can really see how effective it can be, we'll work with you on more specific preparation skills.

For now, we just want you to be *planning* to prepare. We want you to understand, in a basic way, how strongly preparation will play into Ultimate Selling strategies and interactions with prospects.

Let's sum this up visually:

Table 3-1 Simple Preparation Gives You Game Control

Be prepared and...	...you'll be in control of the situation by possessing important knowledge, ready to be used in an instant
Be in control of the situation and...	...your prospect will see the depth of your judgment, your knowledge, your honesty.
When they trust your judgement, your knowledge, your honesty...	...you're no longer a sales person trying to push something on them; you become their trusted advisor
Since you're the trusted advisor...	...you're the person with whom they want to work

That puts you in control of the each step, each interaction, each negotiation...which you will negotiate so that you both win, continually reinforcing their trust in you and your process.

Okay, you see the concept. You're ready to tiptoe in. So what kinds of preparation are we talking about, exactly?

We mean preparing your attitude: to prepare as an Ultimate Sales professional you must make time to fully develop your sales skills and business knowledge. It means accepting that your sales success is up to *you.*

We mean practicing for a better approach: it means learning to listen to your prospective customers more, and talking less, way less. You have two ears and one mouth. Use them in that ratio or higher.

We mean intentional preparation for every interactions with prospects: thinking about emails you're sending, planning phone calls with a specific objective in mind, being ready for contingencies in face-to-face meetings. We'll show you specifically how to do so in Chapter 15, when we go over Intentional Interaction Planning.

We mean knowing who you need to have those interactions with, by knowing everything you can that's appropriate about your prospect. We'll touch on that briefly here, and then again in Chapter 14.

We mean what we referenced earlier: preparing by knowing what it takes to close your typical sale. We'll have you sketch that out briefly below, and then we'll hit the pavement for it in Chapter 13.

We mean knowing what your resources are (in football terms, who can run, who can catch, and the value of a good helmet!) and being ready to know how to use your resources to support you or at least not weaken you in the game. We mean preparing to stand up to your competitors in every possible and ethical way, as explored more fully in Chapter 12.

Preparation takes time and effort. I'm telling you that all your effort, work and commitment to preparation will pay off in spades. We'll do a little exercise to get limbered up, and then move on to some really fascinating stuff. Please pay close attention and do these quick exercises well. We'll call this exercise, ***Your Selling Story:***

- **What do you sell?** Besides the products or services, what else is generally included or important to prospects? Don't think from your perspective. What do you sell that your prospects need and want? Safety? Peace of mind? Prestige? How will it improve their business and profits? Write it down.

- **What's your sales process?** By that, we mean, what are the steps it typically takes for you to make a sale?
  - Please list them from start to finish. Just jot them down quickly now and save your notes for later, as we will expand on them in Chapter 13, The Selling Process
- **Who's your competition?**
  - Please make a quick list of all the companies with whom you compete. List the products or services that you compete against.
  - Make a short list of your product/services strengths and weakness versus theirs. How do they matter to your customer? (Just quick here, you can go into detail later.)
  - Make a short list of your corporate strengths and weaknesses versus theirs. Great customer response, better financial operations, poor cash flow, etc.
- **How do you find out important information about your prospect?**
  - How do you find out who buys, who stalls, who has the money, who influences them? List the questions you ask, or want to but are hesitant to ask.

The completion of these exercises is critical. Doing them will give you some ideas right away about improving your sales. It'll also set you up nicely for the following several chapters. With simple, deliberate practice and a steady pace, you will be astonished how simple this can be. It will get easier and easier. Now, get out your pen and paper or your computer and get to work.

Once you're finished set them aside for later, and it's on to the next chapter, where we introduce how to Qualify prospects...but as an Ultimate Seller.

Chapter 4

Qualifying and Negotiating

Unfortunately, all too often in the minds of many salespeople, being qualified means:

- They have interest in what I sell
- They need what I have
- They know it costs something
- They called us
- I have an appointment
- They have money
- I have a personal contact/friend at the company
- I believe I can close the sale
- It's a referral
- They have a proposal from us
- We received their RFP, RFI, RFQ
- They sourced us via our web site

Those things do not mean they are qualified using the Ultimate Selling criteria!

We've talked a lot in the opening sections about the probability of success. If there was a ten percent chance of rain, would you take an umbrella? No! Then what's the point of doing work and submitting a proposal with an equally low chance of success?

Most salespeople get a list of the buyer's requirements, come up with a rough idea of the project scope, and then are pressured—either by the prospect or the sales manager—to present a proposal and start talking prices. At that point, all kinds of concerns and objections come up, the deal falters, you try to negotiate, and things all too often fall apart—and you find out the hard, time-consuming and expensive way that the prospect was never really committed to the project in the first place! How many times has that happened? To most salespeople, too often.

The simple brilliance of Ultimate Selling is to turn that process completely around.

In Ultimate Selling, you are to *really* identify your prospect's desires. You negotiate all options—concerns, objections, issues and obstacles—before you ever submit what is commonly referred to as "the proposal." In fact, there never *will* be a proposal in the old sense. You'll give them a "proposal" in the confirmation step—but you'll control the qualifying and negotiating process so that they have written it for you, putting in everything they need and want.

Wait a minute: wouldn't that slow the sales process down? Believe it or not, you really can increase your sales by slowing everything down, listening longer, and having your prospect put their "BUT WE NEED THIS NOW!" mentality on hold.

In so doing, they will discover that they really want to work with you. It's about you getting them more involved together in defining their needs and solutions and delaying the proposal until far, far later in the process. Remember, a prospect will always do what they believe is in <u>their best interest</u>. With Ultimate Selling, you'll help them so effectively that they'll quickly discover that working with *you* is in their best interest.

The first thing we are going to teach you is how to do that: how to start slowing things down while you simultaneously start really qualifying the prospect.

They'll tell you how important it is that they move quickly, that they are committed to the project, that they really need a solution. Believing that with all their hearts, their job is to get you to give them a demo and proposal with costs as quickly as possible.

But we're going to fire them from that job.

Their new "job" is to engage in dialogue with you, and tell you as much as you need to know about them and their needs.

And that's all.

There's no pressure! No pressure allowed from them to get a demo scheduled... no pressure on you to produce a quote or proposal... no pressure to assure them that you're the best candidate or the best company or the best whatever.

So what's our mechanism? Just conversation. Them talking, you listening, asking questions....but conversation that you direct via calm control, without seeming to do so.

You will find that they are actually excited to finally talk to someone who ignores their request for speedy proposals and demos, and really listens... and doesn't seem intent on just selling them something.

Yes. It seems to not make sense. But it works.

To become qualified (and again, long, long before any "proposal" or demo) they are going to tell you:

- when they need to deploy or take delivery of your product or service
- how they will pay for it
- who are the consultants they may be retaining, and their purpose
- who your competitors are
- who the decision makers are, and who are influencers and/or recommenders
- how the buying decision will be made
- what the deal-breakers are
- where you rank and your potential for success
- what the politics are
- whether you're really in the game or not

...and yes, all of this *before* you present <u>any</u> solution to them, of any kind, in any way.

Why are they going to tell you this?

Because you're going to ask them.

It's that easy.

In Ultimate Selling it's just about asking all the right questions—and their richly informative answers will follow. You may not get the answer you want; but it will save you time and energy to find out if they're *not* a qualified prospect far earlier in the game than you've been used to doing. No surprises. No "lost sales". No dashed hopes. Just a calm assessment *on your part* that they are not a solid prospect for you to continue with.

Not only do you want to establish early whether they are or aren't qualified, but we also want you to deepen the qualifying even more. Only once we have the answers to all these questions are they Ultimate-Selling qualified.

Why? To learn what needs to be negotiated, and to set you up for very successful negotiations. Together in conversation you can address any issues with your products, delivery, services, support, terms or conditions, competition or incumbent vendors....and work together to create mutually acceptable solutions. Again, *before* you present any proposal; those negotiated solutions will become part of the proposal later. That's what we mean when we say that they will "write the proposal for you". We'll go over more how to do this in later chapters.

We are going to sum up all questions as falling under one of three kinds of qualifying and/or negotiating topics:

1. Interest: Are they seriously interested or aren't they, and why?
2. Money: Do they have it, can they get it, or not?
3. Decision: Who decides the Yes/No; and how?

Qualified, As Defined in Ultimate Selling Means No Questions Remain

Being a qualified prospect opportunity means first and foremost that there is no "gray" in any of those areas of Interest, Money or Decision.

We must convert any gray into a black or white situation...as a qualified, negotiated prospect/opportunity, or a *dis*qualified one. Please note that an "unqualified" prospect means that they (and you) haven't learned enough to earn the right to move forward any farther in the sales process. And, as you continue interacting, they might or might not qualify, in which case they're still a lead or a suspect.

"Disqualified" means you will not be working with them any more. While they may be perfectly nice people, they haven't earned the right to more of your sales time or business resources.

You might think that sometimes there are legitimate gray areas. If I were to agree with your premise, and approve you allowing it to remain gray, we'd find that sometimes they'd buy and sometimes they wouldn't. So leaving them in the gray area provides you with only a 50/50 chance of success... at best!

If you're going to determine where best to invest your sales time and resources, then doesn't it make more sense for *you* to control the process, and negotiate any and all gray areas into black-and-white conditions? That is a critically required (and learned-over-time) skill of an Ultimate Selling professional. It is also how *we* define a qualified prospect.

How do we find gray areas on Interest, Money or Decision? Let's look at some examples of qualifying and see how quickly they stand out.

Example 1

- They have a definite interest in acquiring your product or service
- There is a definite benefit, ROI etc.
- They have a process in place to decide and procure your product or service
- They don't have any money!

  So the gray area seems to be Money. Can it be negotiated (financing, credit problems, or do they really have it but are unwilling to spend? Or is it real enough a problem that it disqualifies them? Keep qualifying them by asking them and negotiating if appropriate.)

Example 2

- They have an interest in acquiring your product or service
- They have money and are willing to spend/invest it
- They have NOT defined the benefit, problem solved, ROI, etc. (ISSUE!)
- They have not established when your product or service must be in place (ISSUE!)

The gray areas here include Interest (they say they are interested, but it's superficial. You need to guide them to a deep connection to their problems and why they want to solve them, and show clearly the benefits and how they'll recoup the money.

The other gray area is a timeline to buy, so they're also gray on Decision. Once you've done the work above and established a real ROI, you next need to negotiate a commitment to act and establish a timeline.

Example 3

- They have a definite interest in acquiring your product or service
- They have money and are willing to spend/invest it

- They have defined the benefit, problem solved, ROI, etc.
- They have established when your product or service must be operational
- They have a process in place to decide and procure your product or service
- They are considering multiple vendors (ISSUE!)

  How can you negotiate so that you become the vendor of choice in their eyes? This is going to require you returning more deeply to qualifying on Interest.

  Knowing your product or service inside and out and your company in comparison to the other vendors would be extremely valuable here. Equally valuable would be Ultimate Seller skills in how you address that gray area.

  What will you offer to help with that question, to resolve their concerns in a win/win way? How do you position yourself as better than other vendors? Are any of the other vendors somehow connected, like a brother-in-law or one of their large customers? If so, can you make a sale? Maybe not!

  We'll get to the negotiation-skills part, but you can see: it's clear where the gray areas are, isn't it? Can you think of your own examples? Think about the last time you lost a sale and why you lost it. What was the gray area that you left unresolved?

If they *have money*, can *make a decision* but *don't really have a need* for your product or service, you can still sometimes make a sale...but in these circumstances, rest assured it will be quickly cancelled, reversed or be one regretted by the prospect/buyer for a long time. Why? They are deficient in genuine need, so they have no real interest—and it's highly unlikely they will lay any fault for the purchase on themselves.

Be clear that you will have to get comfortable asking direct and specific questions to get the answers to these gray areas, while offering solutions to keep building your position as trusted advisor and to keep control of the play. We're going to work on that skill all through the book, rather than just one chapter. But you can see immediately how valuable it is to get skilled in asking those types of questions. Right away, doing so will give you a stronger sense of control and a better understanding of where you are (and aren't!) with the prospect and a possible sale, and it will actually feel more comfortable to your prospect to be well-managed in this way.

Ultimate Selling has a strict definition of what a qualified prospect/opportunity is. If you're thinking that that they are 90% qualified, it's like saying someone is 90% engaged. Someone either is engaged and intending to marry or they're not.

We want you to keep qualifying past the 90% mark to the 100% mark. You never move forward to the next step, Confirmation, until they are 100% qualified. It may happen by accident. You may have thought, especially as you use this approach, that they were 100% qualified. And things can change. But we want to teach you ***not to rush the process of qualifying*** to get to Confirmation. Someone will pay every time that happens... and it'll be you.

Think of qualification as a light switch—it is either on (Qualified) or it's off (not yet Qualified, or Disqualified). So in Ultimate Selling, we define a qualified prospect and opportunity as follows:

1) **Interest:** They have a clearly defined need, requirement or want that can be fulfilled by your product or service.
2) **Money:** It's crystal clear how they will pay for it and they are willing to do so. All options have been discussed in terms of return on investment and agreements are in place about them.
3) **Decision:** The people who will make the final decision are identified and available, there is a decision-making process in place, timelines have been established, and you have commitment to act on those timelines by all the required decision makers, recommenders and influencers.

Any thing short of these critical requirements cannot be categorized as a qualified prospect or opportunity. They may be a possibility, maybe a strong one, a suspect, lead or future, but they most definitely are not a qualified prospect in the eyes of an Ultimate Sales professional.

We devote a chapter to each of these Qualifying criteria so that with you we can explore various nuances and ways to convert gray into black and white. We will provide ideas and guidelines to assist you in learning how to have them become far, far more than a traditional sales 'prospect'.

In Ultimate Selling, the word 'prospect' is now synonymous with being a qualified prospect. Remember, we speak 'yes or no' 'black or white' "one or the other'...not 'gray'.

In Ultimate Selling, anyone in your current SFA or CRM database, lead list, pipeline and/or sales funnel can *only* be classified in one of three ways. (Refer to Figure 4-1 on page 33) They are either:

A Suspect

Anyone who can be classified or identified as a potential user of your product or service. They may be an existing customer who has not yet been qualified on a new opportunity; a list purchased by marketing or your sales manager; a customer referral, a call-in lead, a lead from your web site, telemarketing group, or anyone else that you consider part of your sales pipeline—and have not yet qualified or disqualified. Until then, they are a cool to warm suspect in your database.

We'll talk about how you 'qualify' a prospect as an Ultimate Seller in the next few chapters, and we'll go over what constitutes our funnel components (Suspects, Futures and Qualified). But before we do, we would like to anticipate one question.

Each company has a way of tracking contacts, whether sophisticated or simple. Moving contacts forward in your tracking system is simple. What to do with the ones that are not moving forward? What do you do about contacts and opportunities that you decide don't have enough interest, money or decision-making ability to merit your time and resources?

Well, things change. Companies that do not have enough money one year find it the next. A decision-maker who was disinterested leaves, and the new manager is all about improvement.

We recommend that you have a way of identifying disqualified prospects in such a way that you know in the future if they are worth revisiting (a tickler file) or if they are of no future interest (permanently disqualified).

A Future

These are partially-qualified prospects who have the potential of becoming fully qualified prospect, but an existing circumstance won't allow them until sometime in the future. They may have an existing contract which has to run out, or inability or unwillingness to fund any termination fee/s. To be a real Future, there must be an agreed—to mutual expectation and dates established to act upon in the future. (i.e. Their contract expires on (date), or their capital budget has been put on hold until (date).

They are **WARM** prospects!

Figure 4-1 The Ultimate Selling Sales Funnel

A Prospect, as in Qualified

They have met all three of the critical elements of Interest, Money and Decision. Clear, straight-forward expectations of next steps have been agreed to by all involved parties.

They are **HOT**!

To summarize, the central process of Ultimate Selling is to move suspects through a rigorous qualification process in order to identify qualified prospects that have a high probability of becoming a sale. It's essential that the prospect earns your investment of selling time and effort. Only then, can they proceed to the Future or Qualified Prospect stage of the sales funnel.

One final note: although rare, it's possible for someone who qualifies to be reclassified as a Future or maybe go back to being a suspect. This can happen, for instance, when things change unexpectedly. Uncontrollable events such as funding for an initiative being cut due to an earnings statement, new management taking over, they're acquired by another company, etc. The chapter on forecasting will discuss such instances in terms of sales revenue or booking projections. For now, re-classify them as a Future or Suspect and schedule a follow-up date with the contact at some later to see if they can now qualify as a prospect. But remember, this is a *rare* instance.... it's not your sales funnel in disorganized flux!

Critical to Ultimate Selling are action and motion. Opportunities need to keep moving from suspects to qualified prospects to an outcome: a sale or a disqualified prospect.

Any time they become passive, it requires reassessment of where they actually belong, both in terms of your funnel and eventually your sales forecast. This is covered in more detail in Chapter 17.

As you work to become a master of a strong, up-front qualification process, you need to look ahead for your desired outcome. Stephen Covey said it another way: "begin with the end in mind." Ultimate Selling is like the NASA space shuttle: it's front-end loaded and requires lots of controlled energy to create liftoff, and then it soars.

That's what you can expect from Ultimate Selling. Let's begin with the end in mind, use controlled energy and knowledge to anticipate and discover the outcome early, and move to it with confidence.

What Must I Avoid As An Ultimate Seller?

An Ultimate Seller doesn't get caught in the trap of handling options too late in the game. An Ultimate Seller doesn't rush to present a proposal, full of options to consider and discuss. You never, never, ***never*** want to be discussing options and choices at signing time, at Confirmation. All that gets straightened out *upfront* during Qualifying and Negotiating.

We've said this several times already. We repeat it becasue it runs counter to what people typically do, and it's a tough habit to change. You may be surprised to learn that most if not all indecisions or delays at the traditional close are actually created and allowed by the salesperson. It's a direct outcome of bad strategy in pursuit of a sale.

Let me cite the standard, confusing example from traditional selling and then compare it to Ultimate Selling:

Traditional Sales: You present your proposal or solution with options and alternatives to be decided on, thinking you can accelerate the sales process.

Traditional Sales Result: You have created a circumstance where they're incapable of making a Yes or No decision, because you've offered it as "do they want it with Option A, or Alternative B?" Oh no! Variables popping up all over the place have caused you to totally lose control of the sales process. Get the picture?

Saying "No problem, I'll just add it to the proposal as an option" is something an Ultimate Seller would strive very hard to never let happen. Since you're going to have to deal with it anyway, you want it up front, decided and done, negotiated long before it's time to confirm...not screwing things up at the end.

Dealing with options earlier rather than later in the sales process, you must also *justify* all options that are to be included in the solution. If you and the propspect can't provide justification for an option—it offers cost savings, personnel relief, operations streamlining, maintenance improvement, something—that option is excluded and will not be part of the proposal.

You might be saying to yourself that the prospect insist that you operate a certain way if you want to submit a proposal or be in the game. Well, an Ultimate Seller has rules too. We never do it the "oops, we're at the end of the road and out of options...now what?" way.

Here's one of our rules: If you're an expert, you're interested in serving them and you're experienced. Because you're interested in truly serving them and you bring experience to the table, you're valuable to them. Your time is valuable to both them and yourself. If they're not looking for that in their salespeople, then there's a hidden agenda or motive. Stand strong: this will help you find it out!

In my experience, I continually see salespeople and organizations scrambling helter-skelter on the back end of the sales process to make something happen. In most cases, they're usually correcting a mistake or mistakes that should have been addressed earlier. It usually does not result in the desired outcome. And when it does, the terms and conditions are such that you may do business…but it's not good business! What did the prospect have to give up? What did you? These things become problems in the future. Stick with a controlled, calm win/win, and you will avoid both immediate and future problems.

Here's your challenge. If you think that simply increasing your pipeline will result in more sales, you will hate Ultimate Selling. Go ahead and run yourself ragged chasing more and more unqualified leads and opportunities. We wish you well.

However, if you are convinced by what we've shown you so far about the Ultimate Selling methodology and are interested in knowing more, then I believe you will experience a greater return on your investment, while creating superb customer relationships that are both profitable and sustainable.

Down To Doing It: How Do I Qualify Someone?

While your primary objective is a successful sale, we need you to stop thinking of yourself as a salesperson, locked into the traditional selling methodology of 'Prospect, Qualify, Propose, Close'...or Prospect, Qualify, Propose….Lose. Right now, start thinking of yourself as an 'Ultimate Seller' and win.

Qualification begins with your very first interaction and continues throughout the entire sales process, regardless of whether the communication is an onsite meeting, phone call, text, internet meeting, email, etc.

As quickly as possible when you engage with the prospective customer you need to shift your focus from initial bonding and rapport to a fact-finding mission designed to completely qualify the opportunity, and to uncover and negotiate all issues.

Although every sales process is different, in most cases the qualification process will involve *many* interactions about what your product or service offers versus what they need, and subsequent negotiation. In each interaction, you will continue to segue very quickly from friendly opening conversation to fact-finding.

We call this 'intentional interactions'. The process is learned. We'll deepen the concept about it in Chapter Fifteen. For now, we'll stay focused on learning qualifying and negotiating on a high level.

Let's continue exploring Qualifying by tackling the three main considerations: Interest, Money and Decision.

Chapter 5

Qualifying on Interest and Getting to Their Real Reasons for Buying

In general, it's harder to qualify on Interest than on anything else. The idea behind Interest questions that you'll be exploring with your prospects are *what's affecting them? What stress is it causing to their company? What do they want to gain? What do they want to change? How might that change?, challenge you?*

Their interest, basically, is why they'll make the decision (not how). A company may be able to find money where there appeared to be none, and they may be able to create a decision-making process on the fly, but if they don't have enough interest, even with all the money and decision makers in the world, they aren't likely to buy anything.

Are they are struggling and need competitive advantage? Are they being proactive? Is this purchase because they want to "outdo the Jones"? Does it allow HR reductions? There are many possible reasons, and you need to know what motivates them.

While many of the reasons will be based on logic and in fact *most* of the reasons initially given will be based on business logic, it's the underlying emotional reasons that often cause people to choose one path or another...and prospects are no different.

They called you or answered your advertising or took your cold call or filled out a questionnaire on your website for a reason. Qualifying on interest means finding out what the reasons are.

It's not because they are craving to spend money. It's not because they have a great team to run projects. It's because something is affecting them enough that they want to take action.

What is that thing? We need to talk to them and find it out.

We're talking about business-to-business sales at this point. These typically are not impulse buys, as retail shopping or other business-to-individuals sales can often be. Their business may have contacted you on impulse, but generally it's because something has been gnawing about them for a while. If you sell to individuals, you can apply these qualifying principles, but you'll need to adjust for your type of sale.

Back to the prospective customer's interest: you've called them in response to a web hit. After some pleasantries, you'll want to start finding out what they are interested in buying, and more importantly right now, why did they contact you?.

They will buy because:

- It saves them from losing money, or it makes them more money
- It saves them heartache/time/hassle,
- It makes their business more attractive
- It rescues or improves their status (in eyes of consumers, competitors, stockholders, employees, themselves)

Buying has a definite emotional component, and change can be both exciting and fun. It can feel like progress, even if it's the wrong thing entirely! An ethical seller stays focused on asking the right questions to make sure that the sale is a win for the prospect as well as for the selling company.

NOTE: *Keep making sure you're talking to decision-makers and influencers!!!! If the Decision-Makers won't even meet with you, they're never going to have enough interest to purchase from your company. And do you want to be hearing the emotional issues of someone who may be a nice person and stressed about their job, but who really is venting at your time and expense? Chapter Seven addresses getting to decision-makers.*

Besides clarifying the reasons why they are thinking of buying, qualifying on Interest is also where you will start defining these reasons to prepare for qualifying on Money. Qualifying on Interest identifies what is the *emotional* cost/status of the current situation Qualifying on Money identifies what is the *actual* cost of the current status, versus the solution. We'll go over this more in the next chapter, Qualifying on Money.

No matter what you sell, defining interest means learning to find out what their current status is, how they feel about it, where they want to be, and how they think they will feel about that.

As you qualify this way, you will immediately start changing your focus from selling to listening. You'll also have an extremely professional demeanor. You'll be calm. It will be reassuring to your prospect and they will be encouraged by the fact that you are listening to them.

Whatever your personality, being professional means remembering what we said earlier: as quickly as possible each time that you interact with a prospective customer you need to shift your focus from initial bonding and rapport to specific, direct questions...in this case, about their reason(s) for buying.

Please be aware that while this chapter focuses on Interest, the subject of Money and Decision will also surface. That's okay. There is never a straight-line process in which you dust off your hands and announce, "great, we've qualified them on Interest. Now let's start on Money!" In truth, you will be constantly engaging them on all subjects. Sometimes it will be only about money, but very often, the subjects of Interest, Money and Decision are very much entwined.

Go with it! Our goal is to compete all of the questions in the chapters on Qualifying for Interest, Money and Decision in whatever order they need to happen, to determine if our prospective customer is truly qualified for sales pursuit. We will accomplish these intelligent, insightful questions both via planned, intentional interactions, and some game-time calls.

You'll find that each area of qualifying has its own characteristics—and it's helpful if you can be flexible enough to manage them. Qualifying on Money tends to be a fairly cut-and-dried procedure. It requires ability of following up on details and paperwork. Be prepared for this or get someone to help!

Qualifying on Decision tends to be about following processes already in place (with their company) and sometimes being a bit of a go-between for different parties. You'll need to be ready to create solutions. This may require the most creativity and mediation skills of the three.

Learning about Interest, however, can be and often is based on emotional issues—frustration, loss, fear, desire to succeed, excitement about possibilities. Qualifying on Interest can take you into all kinds of emotional territories! It helps if you can be comfortable with some level of emotions in the conversation. There may be real fear of losing staff members or their own job. They may be feeling fear or anger or frustration at a competitor who may take business from them, and the need to fight them off. There may be the desire to be home with their family more which they are looking at your product to somehow solve. There may be impossible environmental issues to address. The list is infinitely long.

The balance we recommend that you strike is one of an attentive and active listener, but not too familiar. Remember, you're there to sell a product and be professional, not become their best friend or therapist. (Even if you do a stellar job with the product, and their goals are attained, don't look for the kind of recognition that friends get. It's business!)

So stay focused, no matter what comes up in Interest qualifying, and ask good questions. We've given you lots of examples below. Keep steering the conversation firmly and gently back to the topic at hand and your reason for being there. Stay in control!

While you'll learn to control qualifying on Interest, Money and Decision, it's also important to be flexible enough so that they feel comfortable with you. There's a huge difference between the calm control of leadership which invites interaction, and rigid, autocratic procedures.

As you review the Mind Map in Figure 5-1, you can see a visual reference and drill-down questions that will help you 'peel the onion back' in different areas to get to real reasons. Refer to this Mind Map on Interest as you craft one for your own sales process.

As you talk about their interest, you will likely need to help them crystallize their needs and requirements, even define some of their wish list. As you drill down on the items/options that you offer or they want, negotiation will begin. For the purpose of this chapter, we're going to assume that you completed the previous small exercises. You can speak clearly about the value of your products or services. You know how you stack up against your competitors. This knowledge will help you to frame questions to your prospective customers in order to position yourself for success and uncover any weaknesses.

Negotiation means testing your ability to meet each need/requirement/option/wish against a measurement criteria *that they set for you.* This is where you take a very professional approach in reviewing their stated business reasons to buy (save/make money, reduce negative factors, improve status) and justify the reasons with them, *in their numbers and their words,* to show the return on their investment. At the end of this chapter, we'll go over that again in depth.

Figure 5-1 Qualifying on Interest

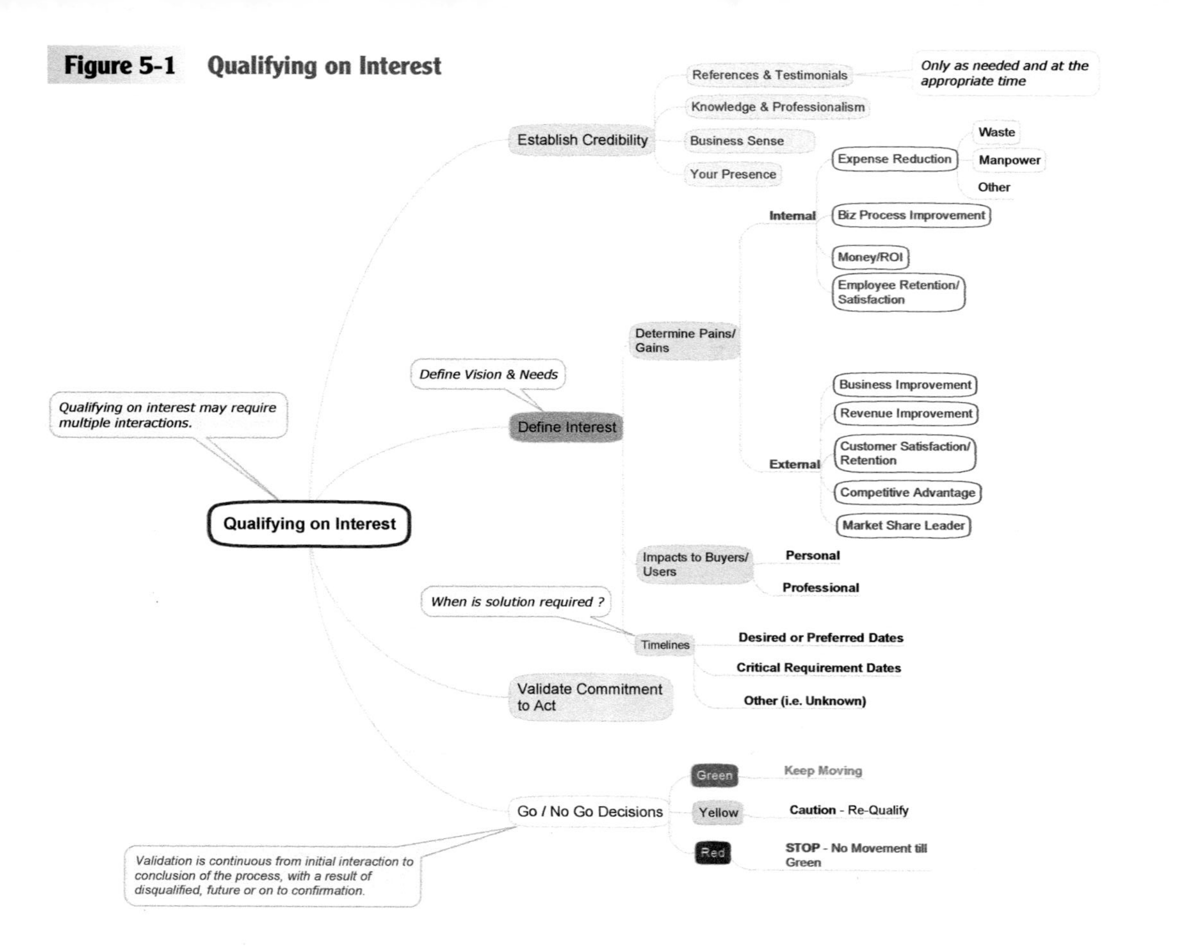

For example, your prospect wants to increase production output by 50% in one month and they seem to have a clear commitment. Let's test it and see if the commitment is legitimate or needs to be revised. You would say something like, "Let's say that we are able to help you increase your production by 50%. It's going to take several meetings to map out the new processes. How much time do you have to devote to this task? Only three hours a week? That means we won't be likely to have an alternate production set up for four months. Is that time you can afford to hold off, or can you find a way to get some time to devote to meetings?"

Negotiation also means uncovering any issues and challenges that would preclude them from doing business with you or hold them back. Do they say your product or service ***won't*** save them hassle? Who told them that? Get your competitive analysis going and help them see—truly—the difference between your services and your competitors. Maybe you offer a ten-year guarantee with free service. Does your competitor? How important is that to them versus price, or ease of use? Don't be afraid to ask,ask,ask simple, direct questions and listen, listen, listen.

Here are some Interest questions to practice with. Ask them politely. Ask them gently. But ask them, very clearly and directly:

- What can they tell you (as much information as possible) about their existing situation? If they are looking at a product or service they've never previously utilized and must now pursue, how will they handle the problems they think the new change will bring? Why are they looking at the new "stuff"? How did they hear about it? What are their expectations with the change? What speed bumps or hurdles do they expect or anticipate during transition?
- What are the various impacts their current situation is having on their business from an operational, legal, and emotional perspective? Are people working too hard and too long because of it? Are they bleeding cash? Ask, ask ask…and remember, ask open-ended questions instead of 'yes-no', and engage in active listening.
- What problems are they experiencing with their external systems (customers, distributors, vendors or market position) from their current situation that they hope your product or service will resolve? Are critical supplies not arriving, and costing them business? Are they unable to get financing because of poor sales performance?
- How would they characterize their level of interest? Are they simply exploring evolving options as the market changes? Have they moved

from 'developing awareness and getting educated' to creating a budget? That means their interest is stronger. This will overlap strongly with qualifying on Money. Focus on the WHY rather than HOW when qualifying on interest and the topic of money will usually come up.

- What's their awareness about other organizations who have tried a similar solution?

In Ultimate Selling, we DO NOT trivialize this process. It's real. It's direct. It's honest and unflinching. Much like having a thorough physical by our doctor, if it's meticulous, it can be reassuring. Like a professional, keep it personal, conversational and as comfortable as possible for the prospect.

Careful, thorough questions give us the opportunity to apply our knowledge, expertise, experience and intuition via intelligent probing. Over time, this process gives not only us but *the prospect* insight into the right solution...becoming one that they helped craft and in which they become deeply invested.

With that said, while you always treat prospective customers with respect and dignity, listen with empathy, relieve the pressure of the decision and create a solution that can work for them and with you, it does not mean you won't take positions and hold firmly to them. It has to be a Win/Win—for you as well as for them.

Let's review some guidelines for Qualifying as follows:

- It's not limited to any timeframe or number of interactions. Depending on your sale, there may need to be only a few or many. It continues until you've been able to fact-find enough to uncover and validate your prospect's needs and desires (pains or gains) in relationship to your product or service offerings.
- It allows the prospective customer to justify financially *and* emotionally the acquisition of your product and service, together with you, so that they feel well managed and recognize that they will be getting what they want.
- It requires a comfortable process and exchange of information for all parties and feels more conversational and less interrogational. Pace your dialog so the prospective customer is never overwhelmed
- While the prospect talks about their goals, you will gradually accustom them to start a negotiating conversation. These will start the

foundation for Money negotiations as well. Build the bridge to your product or services.

- You'll need to become adept at recognizing whether you have a red light, green light, or yellow light—and if you see a yellow or red, slow down, and take time to address those yellow and red lights and turn them green in order to proceed.
- It's where good preparation will really pay off, as you will be able to focus totally on your prospect from their perspective, not on what you will say next. As a result, you not only hear what they say, you will understand what they mean and be able to respond strategically.
- You will demonstrate empathy and put yourself in their shoes so you can see and understand the big picture from their vantage point… like seeing your opponent's hand in a card game, and helping them play it well.
- You don't make any presentation or demo until fully qualified. That's reserved for the Confirmation step.
- When you work with a prospect, it's critical to keep track of what's accomplished, what happens next and after that, and what's to come. Follow up every verbal interaction with an email re-capping what happened and what was agreed upon, or what progress was made, or what you agree should happen next. Make sure you get responses. That simple accountability will set you apart in their mind as being a consummate professional - part of the tenet of an Ultimate Seller.
- You cannot consider the Interest step complete until you have determined their needs both at a superficial level and at a deep, below-the-surface level—so you know *all* the reasons why they are looking. In addition, with the prospect's help, you'll create perceived value that will be greater than any financial investment you will be asking them to make when its time to qualify on Money.

It's okay to determine right here whether to proceed with the sales process, or to close the file or suspend for some future date. You're at the point that you know that they are now either a prospect qualified on Interest, a prospect disqualified on Interest, or they are possibly a Future. There's no gray on Interest. You've worked asking questions and negotiating until its black or white, a clear yes or no on Interest.

We've talked about a lot of things that were important to them as you qualified on Interest. Do you know which are their *biggest* reasons to make the decision? Is improving the reliability of their products most important…

or is it that they don't want the after—hours phone calls any more?" It's generally an emotional reason which can be very powerful to understand.

Ask them to tell you their most important reason "why" to give you powerful insight into what's really driving the project from their personal viewpoint as well as their professional one.

You could even add more questions, such as: ***why us?*** This provides another opportunity to ascertain where you stand and uncover what else you might need to do to ensure the outcome.

- "Why do you think you want to do business with us, Tanya? I know some of our competitors are pretty big firms.
- or even more open-ended, "I know in this economy you've got lots of vendors hungry for a sale like this. What are you looking for in the company you choose?"

They may come back with "low bidder"—which tells you more about Money and Decision, or they may answer something generic like "we want the company that will provide the best value or most after-market service."

That, my friends, is another question begging to be asked. "Thank you, Tanya. In your view, what does great after-market service look like?" Every conversation about their needs and desires gives you better insight about the issues that will form the emotional foundation on which the numbers for their ROI will be plugged.

You will find out if you and your company are way, way out in left field. It's better to know that early on! Knowing where you are versus what they want is powerful.

What do you want to ask?

Let's move on and go over a list of possible questions about interest that can be used with prospective customers. As you will discover, these questions, and asking questions *behind* the questions, should address all pertinent information you need to know to establish their actual level of interest.

The questions are better asked in such a way that your knowledge becomes something intriguing, something that they perceive more and more as being valuable to them. You are looking for them to be thinking, *"They must really know their stuff! They're experts…they ask such great questions!"*

As you review the list below you can see how one question evolves into subsequent questions that dig deeper, that peel the onion back, to get to the real issue or challenges. As you go through the list, identify those that would be most useful to you:

- What is your current situation?
- How long have you lived with this problem?
- Why has it not been addressed sooner?
- What is the impact or payback from fixing this?
- What are the impacts of not fixing this?
- What do you know about our organization? What are you looking for in a service partner?
- What, if anything, has kept the organization from solving this problem before?
- Can you be more specific?
- What have you done to address it?
- How did it work out?
- How critical is the delivery timeline?
- How did you feel about that?
- What else should I know?
- How will you solve the problem if you don't engage us to fix it?
- How has it affected your customers and customer satisfaction?
- How has it impacted your staff?
- How has it impacted your operations?
- Since you have an existing supplier, why am I here?
- What would cause you to end your relationship with them? How would you go about doing that?
- Are we involved for the purpose of keeping your existing supplier honest?
- Do we really have an opportunity to displace them? Why?
- Since you are using a consultant, do you know if the consultant and or their firm has ever recommended or used our product or service?
- Understanding your needs and requirements, the marketplace and other products and services, it's clear to me that we can meet and exceed about 90% of your requirements, albeit the most important ones and not meet about 10%, albeit minor ones. How acceptable would that be to you?

- It's great that you want to use new technology in solving your problems. However, unless my observations are incorrect, you have used little technology in the past to address these issues. What's different now? Why now?
- Typically, we are the highest priced provider and while more than justified in our overall value proposition, how can that be addressed with others involved in the decisions making?

It's almost impossible to list all the questions that will surface as you're interacting with your prospective customer. Therefore, stay focused on the prospect and what they say, and get below the surface to find out what do they *really* mean. Learn what are their intentions, and you will hear and discover the opportunities as they present themselves. This will guide you both with your next questions and open up areas for negotiation. .

Remember, be patient. ***Don't start selling.*** Keep listening. Negotiate on points as needed. Ask more questions. You will have your opportunity to show them the value proposition in the Confirmation step. Save it for then!

It's the Questions Behind the Question

Critical to Ultimate Selling is to understand what their answers *mean*, not what they say. That's why additional questions are required by you to uncover their real issues and challenges…the ones they will really spend money to solve!

Asking great questions continues to reinforce your credibility as a trusted advisor and expert. It demonstrates your understanding of their issues... —and it's how you relieve pressure on them and get to the root issues, not just the symptoms.

Can you imagine a doctor prescribing a solution or medication without first understanding the real problem? I doubt it, as it could be malpractice. In Ultimate Selling, we don't want to commit sales malpractice.

Let's go over some scenarios to help you practice a little more.

Interest Scenario 1:

Prospect: We need delivery by next month!

Traditional Sales: Great!!!

Ultimate Seller: Great. That said, I'm curious as to what happens if it's later than next month?

Interest Scenario 1—Summary: You might ask this question even if you can deliver today or tomorrow. Why? To gain more insight into their need—and to give you an opportunity to begin negotiating, even if you know you can fill their need easily. I guarantee you will know more about how much they really need it and if delivery is an issue. If it is, it could play into a higher margin of profit for you by expediting delivery; it may eliminate others who can't, or may offer other benefits you can present in your value proposition, or help you steer away from negatives you must avoid in presenting your value proposition.

Interest Scenario 2:

Prospect: We need it right away!

Traditional Sales: No problem!!!!

Ultimate Seller: While right-away delivery is not a problem, I'm a little curious if something on your end could change that. You stated earlier that you attempted to purchase twice before and it never happened. Could it happen again?

- If they answer yes, you ask about the probability in percentages.
- If they say 75% likelihood, you would ask, *so why are we rushing to do this now?*
- If they say 10% likelihood, you would ask *how we can eliminate the 10%?*
- If they say No, you would ask *what's different this time*

Interest Scenario 2—Summary: In both cases, you now know more about what's happening and can act accordingly. You used information shared by your prospect in their dialog about prior attempts. You need to determine how firm it is this time or if something could change—and if it could, how interested are *you* in continuing to working with them? Are you convinced it will happen, or not? How high is your probability for success?

Uncover what it might be so you can address it now or disqualify them. You need to predict with a higher level of accuracy what will happen.

If you accepted their first response, as most traditional salespeople do, you'd probably feel safe projecting it as a new sale next month in the sales forecast…but like too many forecasted sales using traditional selling, it wouldn't happen as projected.

Interest Scenario 3

Prospect: Will your system crash?

Traditional Sales: We have a very high Mean Time Between Failure Rate, so it's unlikely this will happen. How soon do you need delivery?

Ultimate Seller: That's a great question. What makes you ask?

Prospect: ABC Computer said that you have system crashes.

Ultimate Seller: Thanks for your honesty. The fact is, they are computers and at some point in time, the software or hardware will crash. Having said that, unlike any of our competitors, our systems are built with redundant processors and our software must pass rigid testing.

So, while it's not impossible, it's highly unlikely. Now, let me ask you, how important is reliability and up-time to you?

Prospect: We cannot risk any down time due to loss of production and customer satisfaction.

Ultimate Seller: I understand. Unlike some of the other manufacturers, the reliability we offer is reflected in our higher pricing. How will you justify this? What flexibility does your budget have?

Prospect: I'm not sure!

Ultimate Seller: What is the financial impact per day or hour if you have down time?

Prospect: We don't have firm figures on that.

Ultimate Seller: No problem. Let's see if you and I can create some realistic estimates of the financial impact. Knowing that will give you a solid foundation of information to set a realistic budget. Make sense?

Interest Scenario 4 (Expanding on Scenario 3)

Prospect: No, I won't provide any numbers or work with you on them. I'll wait to see what's in your proposal.

Ultimate Seller: I'm confused. Does that mean you will make a decision about your purchase without any financial justification for redundant systems, and just hope for the best?

Prospect: That's the path we'll take.

Ultimate Seller: Forgive me, but I'm still confused. Based on what you just said, how can you succeed?

Prospect: That's really not your concern. (Are they being a little adversarial? Yes? Why?)

Ultimate Seller: It sounds as if you may have another vendor selected and want us in the process to keep them honest or there are others involved in the decision-making process we have not yet involved. Either way, don't spare my feelings and tell it to me straight so I can understand and support your efforts. OK?

Prospect: You're correct. I do have an IT vendor I want to keep honest. How can you help me?

Ultimate Seller: I can help you by not wasting any of your time and yet provide you with a way to keep them honest. Interested?

Prospect: Yes, but how can you do that?

Ultimate Seller: Are you the decision maker or are others involved?

Prospect: It's my decision.

Ultimate Seller: Great, here is how I can help:

- We will not submit a proposal, as it a waste of your time.
- Once you have received your other proposal(s), delete any vendor identification and financial information and submit to us for analysis
- We will review and analyze based on the specificity of requirements of the product and/or services offering
- We'll submit our findings for your discretionary use

Prospect: Why should I do that and why would you?

Ultimate Seller: You would get the benefit of expert analysis at no cost to you and we would not invest time in money for a lost cause...but we could benefit from the process and update our current competitive intelligence. It's a win/win without making a sale. Does that work for you?

Prospect: Ok, let's do it!

Interest Scenario 3 and 4 Summary:

- How much did we uncover from this prospect?
- Did we drill down until they ran out of answers?
- Did we uncover additional information?

- Did we ask another question when the answers provided did not make sense?
- Did we uncover their real motives?
- Did we focus on their goal and provide a benefit alternative?
- Were we honest?
- Did we do all the proposal work for another loss?
- Did we propose a win/win alternative?
- If our 'competitive analysis' offer is accepted, do we still have a chance to covert to an opportunity based on findings?
- Is our offer ethical and honest?
- Were the prospect's original intentions ethical and honorable?
- Did we control the process?
- Did we provide stress-free solutions?
- Did we position ourselves as the high priced provider?
- Did we position ourself as a credible expert with much to offer?

Just think: as an Ultimate Seller you were able to accomplish all these things without making a sales pitch! It was conversational, void of traditional sales, focused, to the point and explored alternatives.

As previously discussed, you can see how our dialog in the Interest step touched on Money and Decision but did not focus on them yet. And who controlled the scenario? You did as you learned their intentions early in the process, and can now focus your efforts on other prospects.

Interest Scenario 5

In this scenario I want to address the subject of an RFP (Request for a Proposal) RFQ (Request for Quote) and an RFI (Request for Information) in a narrative format and provide some insights as viewed by an Ultimate Seller

Each of these has their own requirements. In general terms, the RFI is to gather information, the RFQ is more focused on getting a price quote, and the RFP offers specific, detailed solutions to defined needs. It may have been developed internally, by a vendor such as yourself, or by a third party entity, (i.e., a consultant.)

As an Ultimate Seller, your objective is to pursue qualified opportunities that have the best chance of success. If you were not involved in the development of the specifications in a way that could position you as a vendor of choice, then most RFPs would not meet the qualification level. Ultimate Selling is about your ability to control the dialogue, which can be very challenging to do in an RFP situation.

For this discussion, we will focus on the RFP and set aside the RFI and RFQ as you can apply similar principles to them.

Since you did not have any significant impact on the specifications and development of the RFP, it's like being in a card game with the deck of cards stacked strongly in another's favor and much less in yours. I ask you, how many of those types of hands do you want to play? But if you're stuck, here's what you *can* do with an RFP:

- Learn as much as you can about the decision-making process. If it's going to be low bid, decide if it's worth your while. If it's going to be expert-skills-based, find ways, if at all possible, of having the decision-makers become aware of your skills outside the submitted RFP.
- Learn as much as you can about their interest. There may be parts of your product or service that are of particular interest to them, while other features that you think are valuable are not of any interest to them at all. Influencers and gatekeepers may be able to help, as well as the decision-makers, if you can word your questions skillfully enough.
- Use any tactical strategies that are fair, including freedom-of-information access into what's happened in their past buying. Study their website to learn what is important to them and use their terminology.

Some examples of specific questions:

- Since you will be issuing an RFP, will we have an opportunity to work with you on the specifications? If not, why is that? Is it possible to change the RFP development so we (or "vendor input") can be involved?
- Will there be a presentation or any interfacing with the end-user decision makers? Or will it be a straight proposal submission to the consultant and you'll select a vendor?
- Has someone else developed the RFP? Who? Has another vendor or supplier helped them to develop it? If that's the case, why should we respond to the RFP? It sounds as if you've already got a candidate, and you're going through the process for other reasons? Am I correct? What is the RFP process really for?

If and when you can be actively involved in the development of specifications and the creation of the RFP, then you can apply many of the concepts in Ultimate Selling in the process—especially when it comes to money and how the buying determination will be made.

Interest Scenario 6:

A prospective customer calls and requests a proposal for an online (ASP) software program...and you only sell server-based solutions. How to negotiate?

"Please allow me to ask a few questions regarding your application. Have you completed a full analysis of implementing an online solution versus a server-based solution? There are many implications, pro and con, on each type of system. I'll be happy to answer your questions about them.

Or you can expand by continuing as follows:

"Please allow me to be direct. Many times potential customers believe that an online solution is better and then later on discover there are all kinds of issues—inability to get reports, inflexibility of online providers to allow you access to your own data during cash-flow crunches, or slow screen response times which can significantly reduce your operations abilities. In some cases, you could be unable to recover your own data and customer records if you choose another software in the future. Do you have any questions about these things?"

If they are asking for something you know that your product can do, ask anyway. For trucking companies: "I understand that you need containers that hold a ton of liquids, securely. What's been your experience with existing containers with rough handling or load shifting, like hard braking in transport or even and an accident situation?"

If it's something that you know your product can't do, ask anyway also. "I understand via industry research that half-ton containers perform better in all circumstances and outweigh using the larger containers. Is this something we should dicsuss further?

Or, "How open are you about exploring the benefits of using half-ton containers?" (A paradigm shift may be a requirement for any sale to happen!)

You may decide that it's not a win/win because your product just doesn't fit their needs or requirements, or you may have exactly what they want, but need to negotiate on an infinite variety of other issues: sound quality, recyclability, performance specifications, minority representation, fat content, processing speed...whatever the unique properties and characteristics of the products or services you represent.

Last: Having Them Build a Case for ROI, and Setting Up to Qualify on Money

Let's return again to that last important feature of qualifying on Interest: start the process of having them see a return on their investment. As we talk about the *emotional* cost of their current situation—it may be frustrations from lost production, lost sales, returned product, or personnel turnover, or it might be the desire to create positive energy—we'll be creating the story as they understand it. It may be that they are mandated to change as well. In that case, the Interest will be a legal requirement, but there will still be underlying emotions (annoyance, or "it's about time" or "how will we get this done in time?") Becoming a trusted advisor means showing them that you understand about those concerns and will be addressing them with great care.

We don't talk ROI dollars and cents in Interest qualifying. We talk the physical, situational, market and personal ROI at this point.

Later when we talk about Money, we'll work with them to tie these physical, personal, situational, market and legal issues together with the costs they tell you about in their current situation, and put it together to create the ROI.

They will determine those costs because they'll trust it more if the numbers are theirs, not yours. Ask them for their numbers, and then confirm those numbers step by step via email or letters. We said that qualifying is a step-by-step or question-by-question process. Talking about ROI is a perfect segue to qualifying on Money.

You can add to your trusted advisor status in this process by providing great input for them to incorporate. It may be as simple as supplying their own industry standards or ratios acquired via researching of their trade Association. It demonstrates you care, are informed, and are confident.

Let's start exploring how you continue the conversation in our next chapter, **Qualifying on Money.**

Chapter 6

QUALIFYING ON MONEY

You likely have talked about money as you worked at qualifying on Interest. As your prospects come closer to a purchase decision, it is intelligent to help them with accepting the cost and deciding on details of the sale.

Remember that the idea behind Interest questions include "What's affecting them? How? What do they want to change? What will change mean? What will it offer…and how might it challenge them?"

The idea behind Money questions are twofold: do they have it and do they want to spend it? As they are considering your solution, their thoughts should be "*okay, this looks very interesting…how much money does it make sense to spend? What's our payback? Where's the money coming from? Who's the money going to?*"

Will they be able to purchase your product or service, or will they choose to limp along with their current system? A strong ROI with real numbers based on their emotional/logical reasons (from the interest chapter) will help them make that decision.

Where's the money actually coming from? Will they pay directly from operating funds or financing? Will any deposit be required and what are the terms for payment? What steps must be taken to get approval to spend money? What credit requirements exist…contracts, etc? You being calm, deliberate and unafraid to ask here gives them the same sense of your professionalism and a feeling that the project—even at the early stages—is being well managed. That speaks well for you becoming a trusted advisor to them.

Just like your other interactions with them, talking about Money should be intentionally planned and you have a specific objective in mind. When or where it happens in the qualifying process is up to your judgement. It just has to happen!

To initially qualify on Money, you need to have some specific, direct questions. These don't include "how much will it cost". That's always going to be an ROI conversation, which we'll cover below. These questions are the "do they have it or can they get it" type. They do include:

- Are the funds already allocated to the budget?
- Are the funds available now?
- When will they be available?
- What is the source of funds? i.e., where's the money coming from?
- Will it require some form of financing? Leasing?
- Are there any financial issues regarding credit?
- Are there any internal issues that could affect funding?

These questions are not about showing them costs (for the first time) as part of a proposal, and trying to get them all excited to make a buying decision, and then hoping you can get money. They are about helping the prospect plan their financial strategy.

If getting-money issues exist, it's best to deal with them up front and negotiate a solution for them long before you present your value proposition. Trust me, they will be most appreciative of the effort. Openly and calmly exploring money with them will allow them to actually help you build the value proposition that's presented to them in the Confirmation step.

As you ask these questions, you'll begin to learn their 'stage' in the Money process as well. Their *current* Money position might be:

- EXPLORING: "*We want to change our website, but we have no idea of what a new website would cost.*"
- BUDGET APPROVAL: "*We're trying to get a budget approved to replace our material handling equipment in the shipping department. We must have numbers submitted by March to get approval for our next fiscal year.*" Many businesses and especially government agencies need to have a budget approved well in advance of a buying decision for non-emergency items.
- YOU'RE BEING USED: This one they don't generally come to you and say! But if you could read their mind or ask the right questions, you'd hear "*We really love the idea of a new sprinkler system for the golf course. But I need to test if my current vendor pricing is fair.*
- REAL INTENTION: They actually do intend to buy and have not made a vendor decision yet. That's a great place for you to be.

Finding out what their Money situation is should most definitely affect your strategy and your intentional planning for each interaction with them. As an Ultimate Seller, you want to use *every* situation, *every* interaction, to have them look to you for help with decisions—so even if they are just tire-kicking, or if they are using you, it doesn't mean it will be wasted time.

Money qualifying may be where you flush out the fact that no matter how interested they are in the solution, they're not really looking at *you* as the true potential provider. You'll find that out as you ask them more questions, with the intent of either qualifying them on Money or disqualifying them. Once again, better to find out earlier rather than at proposal time!

Figure 6-1 Qualifying on Money

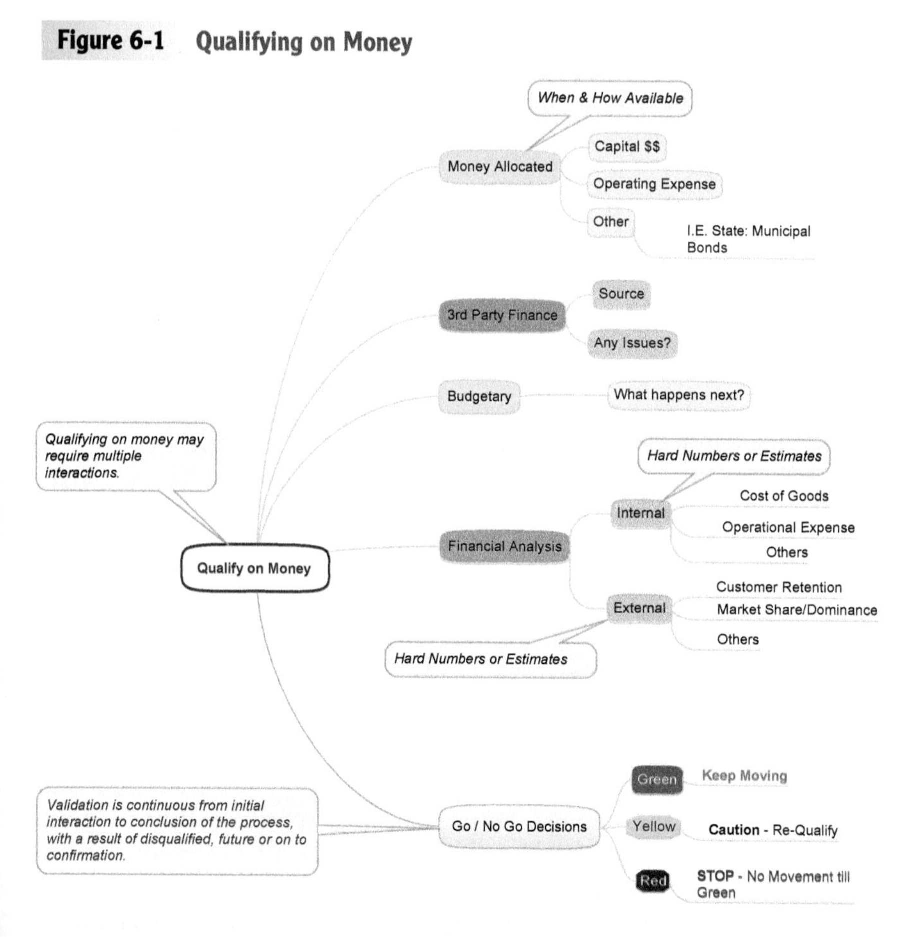

How do we work our way through these scenarios?

One of the most important things, as with all qualifying, is to make certain that you're talking to Decision-makers, plus some Recommenders and Influencers. (The next chapter will cover strategies for getting to these people.)

What if their Money Stage is Exploratory (Tire-Kicking)?

Let's say that they have expressed some interest in your company's ability to manage rental properties for them. Their properties are a hassle to oversee, and they just found out about your service. But when you start talking about Money, it becomes clear that they are really just tire-kicking. Maybe you didn't qualify them thoroughly enough on interest, or maybe money came up in the conversation and you quickly realized that they have nowhere near the money they'll need.

Are they a qualified prospect? No! But this doesn't disqualify them automatically either. In discussing money, you may be able to create interest and a sense of need. Here's where we take what we learned from their interest and talk about money to create a great ROI argument. If you think it's worth the investment of your time (if you think they really may have the money and/or interest can be built) you could strengthen their interest in your solution based on a good ROI.

And the path, of course, is to first strengthen your position as a trusted advisor.

Because of your experience with your product and service, you should be familiar enough with what you sell to have a SWAG (Sophisticated Wild-Ass Guesstimate). If you know something is going to cost between \$20-\$30K, you may say to a prospective customer that, without more details, you estimate that the project would be somewhere between \$30K-\$40K, but it could be possibly be improved if you can work together with them to be more specific on what they need.

Then quickly switch the conversation back to qualifying on interest again. They mentioned that they are losing some business to competitors. How much business? What's the annual cost to them? What ROI would they need and over what timeline, to justify an investment?

As the conversation continues, keep asking very specific questions about interest. Stay focused *on Interest* once you've given them the rough estimate. Unless interest deepens, you have no sale and you'll need to disqualify them and move them out of your funnel. Keep making sure you're taking to decision-makers. Remember, if the DMs aren't interested in talking about money, they're not interested in a solution. You must disqualify them for "No Interest"!

Let's say they *are* qualified on Interest but you determine that they are NOT a DM or Recommender and they can't get you to the DMs or Recommenders. They may have personal interest, but are unable to get buy-in from the key players...but you think the facts suggest they are a possible Future. Treating them respectfully (so that they remember you as they advance in their career) is a smart strategy that takes up very little time and could pay off later.

In these cases, again provide very general ranges which have room for adjustments, and then quickly close the conversation, while using the opportunity to advance your position as a trusted advisor:

"I like your ability to think strategically. What's the likelihood that you could interest your manager or even one of your VPs in our services? If you're able to raise enough interest, I'd be happy to work with you on the acquisition process. Keep me posted."

End the conversation. You've now disqualified the opportunity, but the person may stay in your contact list. It's efficient and clean while being ethical and personable.

Are They Really Just Budget-Setting?

Let's say you have learned that their interest is mostly budgetary. They've told you they need to plan sometime in the future to acquire a product or service such as yours. This tells you that their focus at this point in time isn't about your product or service…it's about how much your product or service typically costs! It's still tire kicking and exploring, but on a somewhat deeper level. So in Ultimate Selling, this budgetary stage results in similar interaction management as Exploratory. In the long run, they may be more interested, or maybe not, and in your company, or maybe not. This means you need to hedge your bets without using too much or too little time.

How to strengthen your position as trusted advisor in this situation?

First, if the people preparing the budget are *not the ones who will be using the product*, tell the budgeteers that you really need to ask some questions of the end users before supplying a number. Think about it: isn't it unrealistic and impossible to provide fairly accurate pricing for some unknown date and time in the future, for a project which has not been fully described?

Help them understand what an uncomfortable position that puts them in! Let them know that you want to offer good input for their continued efforts.

With the decision makers, start over on Interest. Even though their job is to create a budget and get numbers from you as fast as possible, fire them from that job. Instead, give them the job of convincing you that they have a valid interest, i.e. what are their emotional and business reasons?

But you'll probably have to come about it from a money perspective, since they've got their budget deadline smack in the middle of their vision.

Talk about Money in your "trusted-advisor-assisting-with-strategic-planning role". Will they be paying for this from their operating budget or will the purchase be financed? Do they have financing mechanisms, or need special terms.? Are they interested in you sourcing the funding so they do not use their capital?

Make some time to talk *directly* with them (over the phone if not in person, and most definitely not via email) about some ROI. Let's say they're budgeting to change their soft-drink bottles to lower-density plastics. Your containers will meet that need and you have a SWAG figure you can easily give them. But have them negotiate for it: you're offering something of value to them—(a solid budget amount for planning)—so have them give something back to you: time to answer a few questions for you.

Trusted Advisor, considering sales/marketing impact:

"I'm happy to send over some numbers you can submit to your fiscal offices. But I have a question. Are they looking at just cutting production costs? Besides cutting costs, we can help build sales by making it easy to promote "greener" bottles. Who at your company would be interested in talking about that aspect too?"

Since you've spoken about profit and increasing sales, you've deepened the money conversation. Likewise, you can talk about pain—but be careful here to be sincere and compassionate and base it on something they've already established is important to their company.

Trusted Advisor, reflecting on competitors:

"I'm happy to send over some numbers you can submit to your fiscal offices. But I have a question about the temperature control. To the best of my knowledge ***(but you really do know the answer)*** all of our competitors require carefully controlled temperature monitoring before bottle filling, so you will need to include a line item on their estimate for the temperature equipment they require.

"So even though you'll see that our numbers appear higher, the overall cost is much less. Just want to make sure you get a truer picture so your budget doesn't get blindsided by that. *By the way, if you discover that fact is now incorrect I would appreciate you letting me know.*"

You haven't slammed the competition, just spoken calmly about their shortcomings *from a prospect budget standpoint.* You've just controlled the Money conversation and presented yourself as someone who is helping them avoid hidden costs.

As a trusted advisor for a possible future sale but with little immediate assurance of closing it, there are several tactics or strategies for budgetary scenarios:

- You must respond with the absolute minimum amount of information, using the minimum amount of time, that keeps you in the game—assuming you want to remain in the game. Ideally it would be provided in a form letter or extremely abbreviated proposal.
- We strongly recommend that any budgetary numbers you provide be 15–25% above what you suspect your actual price would need to be. This is tricky: you must consider how high you can go and remain in the game. **Having their confidence strongly in hand as a trusted advisor is critical to having them consider your higher numbers as more accurate than lower estimates from competitors from whom they are also requesting numbers.**

Be forewarned that there is a possibility that your information will be shared with another vendor. As an Ultimate Seller, you want to gather competitive information but you don't want to provide it. Keep your letter brief and avoid specific information.

A higher estimate will establish a price point with the prospective customer that you can potentially reduce later in the Confirmation step. It will please them at that time if you are able to show a lower cost or greater ROI. They will be pleased—but this will also protect you should the cost of the project, once fully explored, be higher than you originally submitted.

Likewise, if they share your numbers with a competitive vendor, the competition will be misinformed.

What if they're using you for Price Checking?

Another scenario is one in which they are using you to check prices: either so they can negotiate with another vendor, or get the lowest price for something they've decided they may be purchasing from the internet if you can't match that price.

While price checking may seem to be a very *un*favorable position, it's actually one in which you are wide open to increase their respect for your opinions and sway them towards what you have to offer.

Think about it: how many times have you looked for a bargain—but a good, confident, knowledgeable salesperson demonstrated quality and value to you that changed your mind? The salesperson is a critical, powerful factor in turning price checking into an actual sale.

Again, in talking to whoever approached you, try to make sure as soon as possible that you are talking to the decision makers, not some underling they have appointed to get prices. Remember that it's their job to get a number from you as fast as possible, but you've fired their whole company and staff from that job, and it's now their job to tell you what you need to know to continue qualifying them.

You need to find out if they are price checking, as they will usually just not come up and tell you that. So here's another example of an intentional interaction: make sure you ask them a question at some point early in the conversation, to force the issue (gently and subtly or not, as you choose!). based on the one below.

"If I can offer the lowest price for this service, will that be a reason for buying with me? What if I offer you a slightly higher price, but better service? What if I offer a significantly higher percentage in cost, say 25%, but my product will last years longer?

If you're talking to a DM, you should get a pause and then a thought-out response. If you are talking to a subordinate who has no DM powers, you'll get a blank stare. So you've just qualified a little on decision: who's really making it?

Remember to treat the subordinates or others with respect, however, and recognize them: they are most likely strong influencers! Asking enough questions which they can't answer, in a respectful and intelligent way, will cause the interactions to move to the decision maker.

Then, it's back to qualifying on Interest again! You're right where you want to be, engaging with decision makers, to most efficiently learn if there is a real opportunity for you with this organization.

If the person, DM or otherwise, insists on a hard number and you think they're price checking but you think there's enough possibility to warrant giving them an answer, you have some additional decisions to make.

1. If you absolutely have no alternative, give them a price, whatever you determine it to be, and deal with the result later. *This is our least preferred choice.*
2. Recommend that they get numbers from other vendors and not you. In your thinking it's not a win/win for you to spend time getting numbers together with no conversation about interest, so you opt not to participate at this time.
3. Offer an expert opinion: that once they have received the other proposal(s), they can delete any vendor identification (and pricing information if they choose) and submit to you for an analysis. Your team will review them and submit your findings. They get the benefit of expert analysis at no cost. As the example explained in the Qualifying on Interest chapter, you explain that you may in fact benefit from the process by updating your current competitive intelligence, so it's a win/win. You've created a possibility to still become their trusted advisor and someday vendor. Weigh the time and effort it would take to do this against the opportunity value and desire to get the account.

Prospects with Real Intentions

This opportunity is in the process of becoming qualified on interest, money and decision in a win/win, healthy way that appears to have a good chance of success. You've uncovered some of your prospective customer's interest: from their problems and issues that need resolution, to anticipated gains or their wish list, you know what they want and how much they want it. You've established that they are beyond tire kicking or setting a preliminary budget or price checking: they really are considering a purchase.

This is the second part of money qualifying, where you tie it very tightly to what you learned in Interest financially justify the expenditure based on what you learned in Interest and in so doing, strengthen the likelihood that they will do it with *you*.

Back in Qualifying on Interest, one of your tasks was to determine the emotional and logical planning impacts to their business. Will change allow them to reach a wider customer base or new markets? Will it stop a cash drain? Will it allow them to mitigate other issues? Will it increase employee loyalty and reduce turnover?

Too many sales professionals don't discuss emotional returns on investment with their prospects, and too many also don't discuss financial impacts effectively. I've observed far too many PowerPoint presentations that convey a compelling case for the product or service, and when it arrives to the ROI spreadsheet slide, the numbers are based on the *salesperson's point of view, with the salesperson's assumptions or their own case studies*…and don't include financial information acquired from the prospective customer.

The failure is that it keeps salespeople in a pitching, sales-versus-prospect relationship, or "you versus them"—and they will challenge any numbers from a third party source and question the ROI, so they will be partly qualified on money.

Not so for an Ultimate Seller in which your prospect comes to see the relationship as you + them = they get a better company. For example, let's say your prospect makes parts that they supply to manufacturers. Their spooled-wire products must compete with others both domestic and foreign on price and quality. If you discuss with them if they have the money to purchase your wire-spooling equipment, you'll find out if they can afford to buy your product and upgrade their plant operations , so they will be partly qualified on money.

However…if you discuss with them the financial impact of customer satisfaction and brand loyalty, the improved quality affect on their market position, plus reduced operational costs from spoiled spools and returned products, the conversation becomes deeper—**and their level of trust in your ability to understand their needs and deliver a good solution deepens along with it.**

It doesn't matter if you're selling wire-spooling equipment to parts manufacturers, shrimp to food manufacturers, or carpeting to builders, the clear message **I want to convey is that Money is not just about HOW they will pay for it, but the foundation and justification as to WHY they should**

invest in your solution(s). If you can discuss with them their needs to close more business, improve their customer's outcomes or reduce their operational costs, qualifying on Money can become a powerful tool in your sales process.

It's an important distinction that requires you to encourage the prospective customer to provide you with their business numbers that are related to positive/negative emotions, and then put the two together. We said a moment ago that a prospect will challenge ROI numbers that a salesperson supplies which are created by the salesperson.

If the numbers are yours, they can and will challenge them. When that happens, your credibility is reduced and you're not in the position you want to be in. However, whether they are actual numbers or estimates, the prospect will not dispute numbers they supply to you. When the numbers are theirs, and you've involved them in your process and proposal, you're speaking their language (literally) and you ARE in the position you want to be: on the same side.

Options and Money

Are there options in your product or service that will increase or reduce the price? Nearly all options do so. In Ultimate Selling all options get negotiated as part of the Interest step in the sales process: do they want them and why?

Qualifying on Money more deeply on options follows the same pattern: do they have it and where will it come from? You'll need intentional interactions aimed at **tying the price conversations to ROI based on their expressed Interests. For example:**

"Georgia, there are a couple of training scenarios we can follow. Either we train all of your staff, or we will only your teachers. Now, I can tell you that while our train-the-trainer modules are excellent, they are not as good as instructor-led training. We identified that you want to keep your budget below $1000 per student...but more than likely, you will need to have us back for refresher courses if you just have us train the trainers. I'll draft the proposal whichever way you want...so I need to understand from you whether you hold a minimal training budget, recognizing there may be production errors, or shall we work out a different training budget?

"Felipe, the new cottons we offer are made from long-fiber strands. From what we've projected, they are going to add about $2.30 to the cost of your finished shirts. How's that going to work for you? They last about twice as long in wearing. You identified earlier that you want to market better wear on your products. Do we include the better cotton since it offers a significantly longer wear time?

Or you could offer "Switching to these cottons adds about $2.30 per shirt. Does it make sense to use these gorgeous cottons and mark them higher at retail and market their quality and longer wear, or simply keep your shirts at the $29 price point but advertise better quality and absorb the cost?"

"We talked earlier about the need to have in place by April. But we're going to have to rewrite a portion of the software for you. We'll need to hire a temporary resource to get that done by April. I need to find out which is more critical, keeping the costs down or getting the screens ready by April."

Ask them, ask them, ask them. Do not be hesitant or shy in the least to ask any little thing as it relates to money. It's the most interesting conversation, in business, and should not be rushed.

Like clearly exploring Interest, a clear, calm conversation about Money is direct, real and meticulous. They will feel well managed. It's honest, and will increase their appreciation for your openness, your commitment to good work in addressing it. You've taken another step to becoming a trusted advisor. For yourself and your company, knowing how, when and if you will paid is essential to continuing your pursuit of the sale.

As you continue to mull over how talking about money openly will deepen your relationship with your prospect, let's proceed to the third leg of Qualifying: Decisions. Who makes them and how to get to those people?

Chapter 7

Qualify on Decision: Talking to the Right People About the Right Things

Being qualified on Decision means that you fully understand the answers to four key questions:

Who will make their decision?
When will they make their decision?
How will they make their decision?
Why will they make their decision?

Let's go over them.

WHO will make the decision?

Let's say you thought you were presenting to the decision makers, but you found out at the end of the proposal review that it's going to be forwarded to someone else—with whom you've never talked and didn't even know about. What just happened to your chances of success?

You may have experienced this firsthand. Believing that you could assume the people you're speaking with are the decision makers, or being too shy or uncomfortable to ask them, too often it ends with you saying weakly "Well, thanks, I'll look forward to hearing from [your boss, your board, your consultant…]"

It's a sensitive thing to confront someone with the question "are you the person I should be talking to, or not?" But asking that question allows *them* to move responsibility to the person with the actual power to do something, and allows *you* to not waste time, so it's a win/win.

The only time that people *don't* point you to the right person is if they like to be (possibly inappropriately) in control. There's strategy below on how to find out if that's the case and get past them.

In the meantime, consider: if *you* initiated the contact, then you're in control of the interaction and responsible for not wasting their time. It's correct to ask directly who is the person with whom you should be speaking, whether you are talking to a receptionist, a gatekeeper, or the person in charge of the company.

If *they* initiated that first interaction, you still want to be in control of the interaction and take responsibility for not wasting either their time or yours. So ask them! Ask them professionally, ask them with a tone in your voice that indicates you want to help, but most definitely ask them: "*Besides yourself, who else will be involved in the decision making process? Who will be making the final decision? Who has final authority or veto power?*"

Ask them whatever you need to get the answer:

- "It sounds as if you're doing all the legwork! Who are the actual decision makers? By whom are they influenced? And how?'
- "Is there a board that needs to vote on this? A committee? Who is the committee chair?"
- "Will your IT or finance department be making the decision, or will they leave that in the hands of the users (preferred)?"
- In a group setting. "Who here makes final decisions about purchases such as these? Who elsewhere in your company needs to be involved? Thanks, anyone else?"
- Or let's ask them personally, one on one. "Kim, I know you've got great rapport with your team members. Will you be heading up the decision team about buying the network? What about George? And John? What will their role be?"
- "Elsie, how comfortable are you with making the final decision about the police riot control barriers on your own? What do you think about getting Kylie's input? How about the city council?"
- "I understand you'll be making the decision, Juan. That said, do you wish to have your father's input, or is he strictly retired these days?"

Decide if it makes sense for you to use time to meet with other individuals. Decide if it makes sense to offer to conduct a needs analysis, or set up for them a projected timeline. Ask who should sign off on it. This is another way of checking that you're talking to the right people!

Interact with each decision maker as much as possible in the long or short time you have, so that they can be familiar with you as well. It will help as well when you get to How the decision will be made: are there hidden agendas?

GETTING BLOCKED?

Make it clear that you must talk to the decision makers. Keep politely and firmly asking questions the underlings can't answer. That creates a face-saving transition to the decision makers. If *they* don't have answers, ask politely who you think would take ownership for those questions.

Remember our example in qualifying for Money: "If I can offer the lowest price for this service, will that be a reason for buying with me? What if I offer you a slightly higher price, but better service and overall value? What if I offer a significantly higher percentage in cost, say 25%, but my product will last years longer?"

You'll recall that we said a DM will pause and then answer a considered response, while a non-DM will not be somewhat a loss for what to say. That difference in response is a big clue.

At that point, you want to recognize the non-DM for what they are: "It sounds as if Richard puts a lot of faith in your judgement. If you think it makes sense, let's take the question to him. That will keep both of us moving forward in the right direction. And I need your advice on how he would prefer we present the question—would it be better to meet with the both of you here, get him on the phone or set up a lunch meeting to get acquainted?"

Treating the influencer with respect is obviously a win/win, and positions you with them as a trusted advisor and equal: you're working together to keep the boss happy.

Also from Money, you'll remember that if you keep being stalled on meeting the DMs, it may be they who are blocking it and not their staff. If the DMs won't meet to talk about making a decision that affects their company well-being, game over. They're disqualified.

ROLES AND RESPONSIBILITIES

Regardless of who is involved, you must understand the process and the various roles and responsibility of each player in the decision-making process. Please refer to Figure 7-1, Qualifying on Decision for a visual review.

For example, the technical buyer may play a significant role in the recommendation but is not the decision maker. Possibly, legal or governmental regulations could dictate the requirements to all of the stakeholders regardless of their role in the decision-making process. While that's part of the "how" the decision will be made, it needs to be identified as well as who is affected.

Figure 7-1 Qualifying on Decision

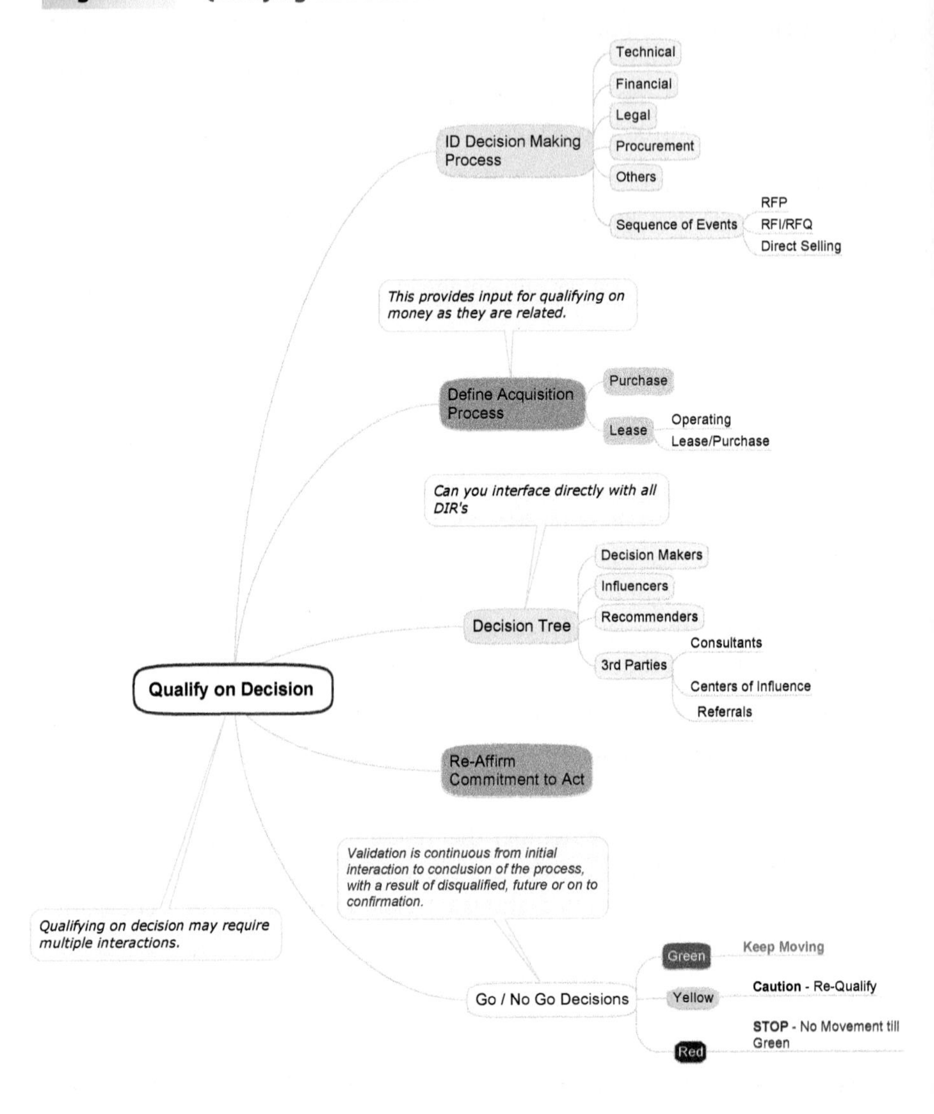

As discussed in earlier chapters, this plays out differently if the sales process is direct selling or RFP/RFI. How would you orchestrate the sales process then, based on the different criteria? You need to align your selling process with how they buy, but that does not mean subordinating the tenets of Ultimate Selling.

The message is clear. Knowing who they all are and the role they will play is critical to an Ultimate Seller. It eliminates surprises and keeps you in control of the sale. Take the time to qualify on who is making the decision.

WHEN will they make their decision?

Many times, they have a timeline for the decision process, and will share it with you. If it's informal, your task is to formalize them: discuss it fully, establish intermediate goals, get it on the calendar and document it.

The absolute best question is to simply ask when is your product or service required for delivery or needs to be operational. From there, you can back into a required decision date and timelines for completing the sales process and executing the purchase agreements. Make sure that they create that timeline *with you*, so that during Confirmation it's their decisions which are being presented.

For example, you might say to them, "I know we've talked about how important this is to your bottom line. If you want to be operational for your upcoming year, you would need to make a final decision by September 15th in order for the project team to start on October 1st. Is that doable?

- or "There's a big campaign on Discover Washington DC coming up this fall. How important is it to you that the new signage be ready by then?"
- or "You've lost two operations managers over the last two years because of production headaches. When do you want the issues to be resolved and provide some relief?"
- or "Since you are sick of the office carpeting and can't wait to have the marble tiling, would you want to use your slow month in December for installation and start the new year with the project finished?

Result: if they have an idea of timeline, you've gotten that part of the qualification started. Your next steps are to negotiate what they need what versus what you can offer. After that, negotiations might go back and forth about money and how it relates to timing options. A big promotion might end that they can't meet, or might be coming up and are advised to wait. That's fine. As long as there is motion forward on the topic, and the When questions are getting checked off, along with the money Hows.

It's not a good result if they *don't* have an idea of timing. Having no ability to give you a firm decision timeline means that no matter how interested they are, and no matter how much money they have, they may never get around to doing anything. You've become a buddy: somebody to whom they can complain, but nothing ever changes.

This is tough. It indicates problems with company leadership. Since you need to control the process, you can take the ball into your own hands and serve as leader, by either taking the opportunity off the table, or by becoming the leader, **if you think it will be successful**. "Jean, we've talked about every possible reason for you to do this, and I know you have the capital. To make it comfortable for you, I'm going to take the decision out of your hands…", followed by:

Option A: you pull back. "I've got several other projects. Let's talk about this next year." Label them as a Future, qualified on Interest and Money but not Decision, and check in later. Be sure to re-qualify on Interest and Money at that time, as things may have changed - but check on current decision makers first!

Option B: you slog through to Confirmation and basically hand-hold them through Connection. Get everything in writing! The problem with accounts like this is that often no one wants to take ownership for making decisions, no one wants to take ownership for good implementation, and likewise, no one wants to take ownership when things go wrong. But they know how to point fingers – so proceed with caution. It's likely to be a headache account. We recommend disqualifying them on Decision and keeping as a Future or Suspect.

HOW they will make their decision?

How they make their decision is not an emotional or financial question, it is a question of process. It may be open to negotiation or it may not.

- Will it be by committee vote? Or by individuals? Low bid?
- Will their consultant be making the recommendation? Will they simply provide oversight, advice…or political cover?

Once you know the 'how' of their decision-making process—or the lack of it—you have what you need to either go through that process with them—or decide that it disqualifies them as a prospect for you.

For instance, if they are lacking a decision-making process (as we discussed earlier) you can choose to disqualify them or you can elect to work with them to create one. Remember, this process has to be an honest open communication like any other, and *you* are in control of deciding if the opportunity offers enough to make it worth both your time and resources.

WHY will they make a decision?

What factors are influencing the decision?

This is where things can get interesting, sometimes. And this is where the lying can happen.

People are not comfortable with telling the truth about some things, such as why they aren't buying. They don't want to hurt your feelings, or they fear that you will argue with any reasons they give and try to talk them into it, so they don't want to give fodder to the cannon or create confrontation.

They also don't want to tell you that they have a vendor that they like better than you, who has convinced them that their company can do a better job than yours.

We don't want you to get trapped here and fall back into the old salesperson versus customer adversarial roles. So we're going to take great care in having you see this from an Ultimate Seller viewpoint. How to be in control and efficient, and keeping things moving?

We started off the book asking why people don't just tell the truth, and it applies here too. Except in cases where telling the truth would get them in trouble with the law (and that's a whole different problem!) why not just tell one another the truth?

It would save a lot of time and expense on both parts. They wouldn't be getting hammered by the salesperson trying to do her or his job, and the salesperson would know that no matter how great his proposal, the even a flawed proposal by a brother in law is likely to triumph, and would not waste valuable time.

Stay with the program. Initiate all previously-avoided, potentially awkward conversations with direct, professional candor. It could sound something like this. "Jim, one of my commitments to you is to very honest about what we can and cannot do for you—even if it means it knocks us out of the sale. I trust that is what you expect from your salesperson, isn't it? Can I expect the same from you? Great! Let's discuss openly where we are. and who's involved in making the final decision.

"Jon, you're normally the perfect candidate for our products. But something has caused you to not move forward. It would be really helpful to me to hear what didn't go right. It's more important to me than not hurting my feelings. I'm new with the company and I really want to succeed, so your honesty matters to me. And don't worry, this isn't a backdoor attempt to change your mind. Let's meet for coffee—you pick, Thursday or Friday – and you can give it to me straight."

We cannot stress enough that these conversations *must* take place verbally, and face-to-face if possible. Although you can email them to let them know you'd like to discuss something, indicate clearly in your email that you'd like to chat about it over the phone or in person, and use the email only to set up a time and place to talk.

We also cannot stress enough that these interactions should be as comfortable as any others to date. Don't be afraid to use humor when you run into a problem. It helps to take off any edge.

Is there someone else, another vendor, with whom they are thinking the money will be better spent? You started the process of asking about competitors as you started qualifying on interest. Don't be afraid to keep asking about it as you work through the process (without sounding anxious!) "We're going to keep working on creating a value proposition for you. Besides ours, how many proposals do you expect to get in and review before making your decision?" Ask!

In learning why your prospects will make their decisions, don't over load them. Be direct but keep it conversational, and keep it very simple—and get those four questions answered!.

Personal ties and political pressure to choose one vendor over another could be tricky. If their sibling's company has the advantage, how can we get them to tell us that? Answer: it's in the earlier qualifying. Go back to the Qualify chapter and work on that some more!

Let's back up and summarize lessons learned about qualifying, negotiating, and being in control of the sales process:

- You must earn and command mutual respect and consideration, or you're just a pawn in the process.
- Negotiate a solution. You must ask direct questions and expect clear answers to the When, How, Who, Why and What questions about how decisions are to be made
- The answers must be a mutual Win/Win…or you must negotiate one to continue the process of Qualifying the prospect
- You will negotiate *in advance* of everything required, to have the conclusion be a simple YES or NO, by not avoiding or deferring any decisions or postponing them to the end
- **Regardless of when, where or how any negotiations or decisions are made was obtained, it is CRITICAL that you document it and confirm in writing** (generally email) your understanding of what you've been told, then hold them and yourself accountable to it. It's extremely important to remind them of agreed-to items and have a trail for accountability on both sides.

Congratulations: they're now either off your list and disqualified for lack of a win/win on decision, and not taking up any more of your time, or they are a qualified prospect. That's all. A qualified prospect. But a very well qualified one!

But they're not a customer. Confirmation has to happen first. And they shouldn't even be on your sales forecast, unless they're within 90 days of finalizing.

HOLD ON: IT'S STILL DEFINITELY NOT TIME TO ASK FOR YES OR NO YET! NEVER ASK for a final Yes or No decision until Confirmation!

…which brings us to our next chapter. Let's review a few points about Negotiating, and then we'll teach you how to get that Yes or No.

Chapter 8

Negotiation Happens All the Time

We hope we've made it absolutely clear that negotiation is an activity that is closely entwined with qualifying, usually happening at the same time, or going from one to the other. In the process of qualifying, we discover issues that need to be negotiated. We negotiate them, we reach a win/win agreement, and our prospect is now "qualified" in that respect. We go back to qualifying on another topic, which creates the next opportunity for negotiating, and so on.

We are saying this just in case it may appear from some of the diagrams, or you may have inadvertently picked up from your reading, that Negotiation is something that happens sequentially once Qualifying is finished. No, that's not the case.

The end of qualifying is reached simultaneously with the conclusion of negotiating. All questions, issues and concerns have been addressed and have had a verbal agreement.

If you're truly a skilled Ultimate Seller, very little, if anything, is left to chance. As a result, once you consider that your prospect is qualified, they should not require anymore negotiation. But if there are *any* gaps to be closed, an Ultimate Seller's approach is one of thoughtful, open discussion of the facts and further negotiation as required.

That's why in Ultimate Selling we focus on 'connecting' being the endgame, not 'closing'. If you find yourself heavily negotiating at the end to somehow pull off the deal, trust us, you lose something: profit, customer confidence, a clean installation...and it can go downhill from there. Thinking less about the close and more about the connection will serve your sales career better over the long run.

There are several key criteria by which you must negotiate:

- You have a strong commitment to win-win rules of engagement. It is not acceptable for either you or the prospect win at the other's loss.

- You're conscious of how your actions and decisions will affect your prospective customer both personally and professionally.
- You're aware that it's much easier to negotiate win/win terms and conditions earlier in the process than later. As decisions are made, they affect other decisions. It's a domino affect! Waiting until later to resolve an item may give you or your prospect less ability to be flexible in other decisions and it can unfortunately force it to not be a win/win.
- You enter each opportunity with a clear picture of where you want to be with price, profit and how to achieve it.
- There are some non-negotiable items: they need the job done right for fair price. You deserve to be fairly paid. You may have to help them understand how that is a value to them. They shouldn't put themselves in the position of needing a heart transplant and then telling the surgeon they'll put it out for the low bidder!
- You understand their willingness to have you as their trusted provider is earned at an emotional level, and you do not take advantage of the trust they have put in you. *If you can't help them, you will tell them so honestly and fairly.*
- Anytime that you have to negotiate, do your homework up front, be prepared and leave little to chance.

In summary, if you end up in having to negotiate more as you reach the end of the sale, you need to understand it's a symptom of needing to improve your sales process and your Ultimate Seller skills. Focus on learning to resolve issues early in the process, and checking back about them as the process continues. Leave nothing to chance or to be resolved or questioned at the end.

Having problems at the end can be because of these problems:

- Were you unable to differentiate yourself from other sellers?
- Unable to establish yourself as an expert and trusted advisor?
- Your product or service was less desirable than your competitors?
- You were not dealing with decision makers?

Wait a minute...wasn't this chapter about negotiation?

Yes, that is the topic of this chapter. We took the long road around to demonstrate that working with your prospective customers to create mutually acceptable conditions early in the sales process is a different type of negotiation. It's much different than "options or price discussions" at the end of the sale—which is what most salespeople *think* is negotiation.

We've already talked about negotiation in Chapter Two, when we introduced the Ultimate Seller approach to qualifying. We also discussed negotiation in the ensuing chapters as we coached you on how to qualify deeply on Interest, Money and Decision. So why a chapter devoted to negotiation here?

Because it's a big change, and big changes bear some repetition.

As we see it you have two choices: negotiate throughout the process, or negotiate at the end and hope for a successful outcome.

As stated many times throughout this book, Ultimate Selling is about deep qualification that goes hand in hand with negotiating. If negotiation happened at Confirming, as it does in traditional sales proposal presentations, and they have the checkbook out, you may find yourself dancing or groveling for the order…an experience the Ultimate Seller despises, and therefore a road less traveled in our world of sales.

If you find that your experience is one of negotiating more intensely as you are almost ready to present your value proposition, you should try and figure out why you end up in that position. Do you actually enjoy sparring? Do you procrastinate resolving some things to the end? Typically, the strategy of deferring important options to the backend of the sales process, figuring you can rush the money and close the deal, backfires most of the time.

It's also a symptom of a bigger problem: your failure to control the sales process. I don't mean controlling the sale with a hammer, or bossing everybody around. Use kid gloves, allowing the prospective customer to discover it's in their best interest to proceed this way, and being confident that they are making these decisions!

(Note: negotiation that happens at the end because it wasn't resolved earlier should never be one of lowering price. If that happens, then the unstated message is clear: you were over charging to begin with. We certainly don't mind a healthy profit—per our example early in the book!—but we recommend that price negotiations take place far earlier in the process.)

With that being said, some buyers are so used to end-game negotiations that sometimes you need to have negotiation points for the prospect so they can feel in control. You should already know what you can draw

Figure 8-1 Negotiate

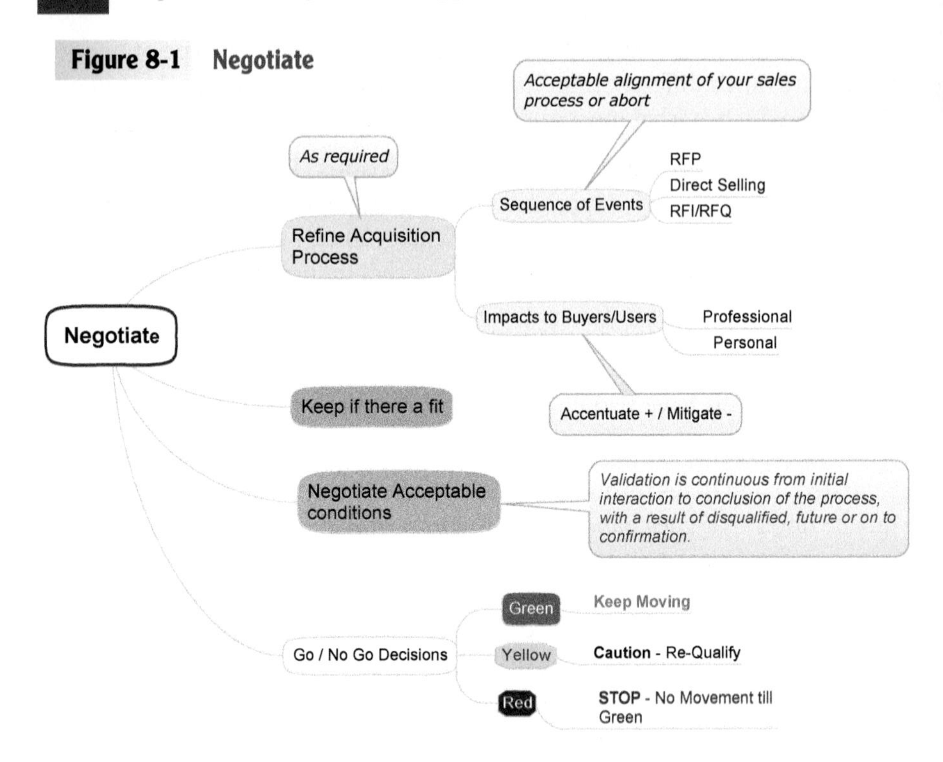

upon from your product or services. These are things that you offer, appearing to negotiate, which are really just give-aways for you. Simply create a scenario in which you can offer them without impacting your price or profit, and enjoy their pleasure at getting the win.

What if they need a win and you have nowhere to go? Simply state that you're not willing to negotiate any further, as it would risk your ability to meet or exceed their requirements. Hold your ground!

Refer to Figure 8-1 as an example of what needs to be negotiated, or not. It will take practice. You'll get better. Watch to see the options discussions start to diminish over time as you get close to Confirming. Watch to see the discussions about how your products or service will help them meet their goals becoming more frequent. You'll sense the shift and see its effect on both your bottom line and your confidence.

They're fully qualified? Everything negotiated? Verbal agreements all around? It's time to go to the next step. Let's talk about Confirming.

Chapter 9

CONFIRMATION

When you began the journey to learn and apply Ultimate Selling to your sales efforts, it's likely that you did not anticipate how different these simple principles could be, and how they would change your communications.

Well, you're almost home. All that's required now is an effective confirmation step. Confirmation sets you up to transition to the ultimate connection: a customer engagement. Confirmation is a specific action in which you provide clear, direct validation to your prospective customer that you will meet or exceed their expectations.

Where Are We Now?

Since you and your prospect have been very thorough in discussing their needs and possible solutions, including all the options and have reached agreement on them, now it is time to offer and review the "proposal". There should be absolutely no decisions left to make...except *yes, we're accepting it,* or rarely, *no.*

Think about a 'no' happening at this point. What would be the reason? While a no may happen far earlier in the game (either by the client or by you deciding to withdraw from the opportunity) if you've really followed Ultimate Selling strategies—having asked all the right questions, and incorporated their desired outcomes correctly in your proposal—why would they say no?

If a no is coming, you've likely have already had lots of warning signals, maybe even some direct comments, to tell you where the problems are. Do you crash and burn at this point? Of course not! You may not have attained the full trusted advisor status you hoped to reach, but you're still in the game. So go back to where the issue is and start over on that area.

We'll coach you on problem resolution in a few moments, but for now, let's assume you've done everything right and it's time for Confirmation.

Let's review some elements to be accepted at the Confirmation stage. They might include:

- Product or service design
- Delivery schedule
- Warranty and support clearly defined
- Pricing
- Training plan
- Installation and implementation plan including timelines
- Financial terms and conditions
- All required documents (contracts, etc.) available
- Special or unique requirements addressed
- Any other items specific to your product or service that were considered essential, and therefore negotiated and included. (Remember that items not considered essential were discussed, negotiated and excluded.)

As you set about qualifying on Interest, Money and Decision as an Ultimate Seller, you defined very specifically the needs and requirements of the prospective customer. In addition, you discussed how it would be paid for, and how the company or customer makes buying decisions.

As you qualified them more deeply, you discussed and negotiated resolutions to their concerns. Those concerns may have been about the value of competitor's offerings versus yours. They may have been about how ongoing customer support would be provided. They may have been about warranties, or finances. They may have had concerns about the need to attract more customers of their own.

If you've completed Qualifying and Negotiating correctly, the prospect has clearly identified to you all the important elements they want in their presentation or proposal.

Let's pause and go through that one more time, and give it a moment to settle in, because you've arrived at the heart and soul of Ultimate Selling.

1. You've asked them a *lot* of questions.
2. You've taken time to listen carefully and make sure you understand their concerns, their needs and their issues.
3. They've told you everything that's important to them, and have financially and/or emotionally justified including it
4. They've basically written the proposal for you.

The Confirmation really is just re-statement (verbal and then written) of everything already discussed, mutually agreed to upon and your confirmation that you can deliver it as stated.

If it becomes anything more than that, something was missed. Remember, concerns, issues or problems that surface at the end of the selling experience become objections, and will put the *customer* in control. Addressing issues early puts ***you*** in control of the experience.

If it happens, you need to revert back to one of the previous steps and figure out where you lost control of the process. Then, learn from it so you can make sure you maintain better control of your process for future opportunities.

So How Specifically Does Confirmation Happen?

There are three critical steps to Confirmation regardless of what proposal format you use (Word, Excel, Powerpoint, etc.) They are:

1. Verbal Restatement
2. Written Presentation
3. The Comfortable Confirmation

Confirm Step One: Verbal Restatement

You begin confirmation by making sure all decision makers are in the conversation together with you. Be clear about that when setting the appointment. If they're not present, reschedule! While there may be influencers and recommenders hearing the conversation, now is the time for action by those who have authority to act. They must be there.

With decision makers assembled, begin by verbally restating all the requirements and objectives the prospective customer set forth: product features, delivery expectations, timing, cost, warranty, whatever. It's critical, before you move to step two, to reconfirm that all those things still apply and that they are still current, stated objectives.

If for any reason you sense objections or problems with restatement, it's because their concerns are still not fully addressed. There is something swimming around in their mind, and they are not ready to make a yes or no buying decision.

Figure 9-1 Confirm

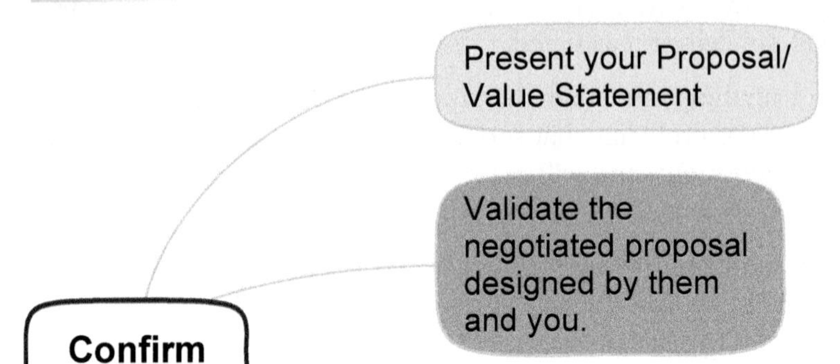

What happened? Good question for you to ask!

If you think back to the Qualifying and Interest instructions, did you take options off the table if they couldn't be justified financially, or did you determine that they are a necessity and have them become part of the understanding? Objections now means they are likely to still be thinking about one or more of those unresolved options. You have NOT positioned yourself successfully as their trusted advisor. You're still a salesperson with something you want them to buy—but they are looking at their options. Not good.

We urge you to slow down and make that sales prospect SOLID by suspending Confirmation. You've got all the decision-makers at the table already. Go back, and discuss concerns, issues, options, until you have a mutually satisfactory resolution.

Here's why: when you want someone to make a Yes or No buying decision, you don't GO to the Yes or No until that's the only, only, *only* thing left for them to say.

Handling Last-Minute Concerns

Your strategy in addressing unexpected concerns (while you are verbally restating all agreements to date) is to pause and ask for clarification on the issue and what changed from what you understood was important. Tell them it's okay, so you can relieve any tension at that moment. Then address the concern: for a few moments, you're back to Negotiating—and get verbal confirmation that the new criteria or resolution is acceptable to everyone.

In a more extreme example, you might ask ask if all of their hard work to get where you are today is for naught, or do they want to resolve the issue and keep moving. The decision to resolve the issue puts you comfortably back into Negotiation again, on that item only.

Negotiation Tip: In order to give something, you should get something. Restated, in order for them to get something at this point, they should give something too.

In a worst case scenario of a no, keep negotiating until you reach a stalemate, when it may be time to walk away.

In any case, stay comfortable, stay conversational, don't rush, don't get nervous and don't back yourself into defending the proposal. **STAY IN NEGOTIATION** until it's resolved, and then you can segue back to Confirmation.

Once everyone has agreed that the re-statement reflects everything that has been discussed and you have a verbal 'yes', you can move to Confirmation Step Two: the written proposal.

Confirm Step Two: Physical Value Statement or Proposal, Presented

It's finally time for an actual written proposal. Again, whatever format you choose doesn't matter, as long as it specifically focuses on and targets all the things required and agreed to, in their language and terminology.

WARNING! *Only use what's required to accomplish this, nothing more and nothing less.* Don't fall into the trap that most salespeople do when a customer indicates that they are ready to buy: don't continue to pontificate, embellishing the proposal verbally. *Know when to stop. Otherwise, it could be the beginning of you undoing the sale.*

The proposal presentation is just that: you presenting all the information that was just verbally agreed upon, but in an actual document and in the format of your choice.

As you review the proposal with the *qualified prospect*, it should reflect how you will address all their identified needs in the Interest, Money and Decision steps. That's why we call it 'Confirmation'. An Ultimate Seller doesn't need to employ the 'trial close' associated with traditional selling.

THE PRESENTATION SHOULD NEVER INCLUDE ANY DIALOGUE ABOUT ACCEPTING IT. Again, this is a re-statement of what you understand their needs to be, and how you plan to address them. If by chance an unexpected, unresolved issue surfaces, (a rarity with Ultimate Sellers) it must be addressed and resolved before you move on. So back to Negotiation, and then resume Confirmation.

Should this situation arise, once you reach agreement on the item, document it. (i.e. Addendum to the agreement, write into the agreement and initial by both parties, etc.)

Once you have completed the review of your value proposition, you may proceed to Step Three of Confirm.

Confirm Step Three: The Comfortable Confirmation

You simply ask all the decision makers in attendance (note: it's not yet time to ask influencers and recommenders who happen to be present—we will get to them shortly) if they are comfortable with how your proposal and Statement of Value will achieves their objectives.

Your goal is to get a solid 'yes' from all DMs involved. Should there be less than a solid yes, I would then ask each person individually the question. At this time, influencers and recommenders would be addressed as well, to flush out exactly who still has an agenda that is not compatible with your winning the sale.

If any lingering concerns surface ***you*** *do not address the concerns! Instead,* you would request decision makers first, and then influencers and recommenders, to respond directly to whoever raised the concern. The rule is simple: ask the person with the highest authority and ability to influence others to speak directly to them. This way, you get to watch and listen to the dialog and learn from it. They will have much more influence on their associates than you. It solidifies the decision maker's 'yes' stance as well.

Based on the Ultimate Selling premise that you did all the hard work up front there should be no surprises and we should hear a resounding, confident, "YES! We are very comfortable!"

That's when we receive our cue that we have Confirmed engagement of a new customer, or new work for an existing client. Congratulatory statements are in order and handshakes all around, followed by the process to formalize the verbal agreement: contracts signed, etc.

Anything short of their wholehearted confirmation is a huge warning flag that something is missing. If that were to happen, you can expect the process to become suspended until the issue is resolved to their satisfaction and works well for all parties concerned—including you.

Remember, in traditional sales, the old uncomfortable, incomplete close was referred to as 'stalls and objections'. Once you've mastered Ultimate Selling in the Qualifying and Negotiation stage, this will not be an issue. Their experience to date of talking comfortably with you and discussing issues openly should have them kick in quickly to help you.

As you work on practicing Ultimate Selling, don't worry if you slip at first, you're learning! Like anything it takes practice to get good at it. The great news is that from now on, your skills will grow faster and faster, as you learn to anticipate and uncover where the problem came from. In your journey of mastering Ultimate Selling, these events will happen less often. Remember, Ultimate Selling is about the pursuit of perfection—which in sales is 100% success on your terms. While you may never achieve a 100% success rate, I can assure you that *pursuit* of perfection will take you much closer to perfection rather than farther from it.

Confirmation is the simple, natural outcome of all your hard work and efforts in Qualifying and Negotiation. CONGRATULATIONS!

It's time to Connect in the next chapter.

Chapter 10

Connecting : The ABC's in Ultimate Selling

Connecting offers you the ability to lay a foundation for a long-term relationship as the Ultimate 'trusted advisor'

In traditional selling, it's really hard to connect with the adversarial focus on "Close, Close, Close!"...even if there really is true value for the prospective customer.

Actually, most traditional methods for handling objections and closing sales can actually *increase* customer resistance—since the objection handling can create pressure and become aggressive, leaving the prospective customers feeling threatened.

Connecting, instead, is very low stress. It's not intimidating and you're fulfilling what has been agreed to, More importantly, it's what they expect. There "no surprises."

New Ideas, And Reviewing Earlier Chapters

In this chapter I will introduce some new ideas and reinforce some presented in earlier chapters. If you've followed the process correctly, you will make the Ultimate Connection, by engaging a new customer.

In Ultimate Selling it should sound something like:

- Now that you're comfortable with the plan, (and this can be asked to more than one party as appropriate) what's the best way to get all the paperwork done?

Point 1: You may even know the answer but by asking, they will restate how it will get done. Remember, no one knows their internal system and processes better, so let them tell you. Then ask if there is anything else you can do to assist them with expediting the process. Again, they will tell you.

Point 2: Now you can manage everyone's accountability, based on their statements and commitments

Point 3: You did not sell them anything; they acquired it from you! It's a subtle yet significant difference.

You have connected because you negotiated win/win resolutions on every issue that surfaced in the sales process, then confirmed it with them. You connected because you have proven they can trust you.

Unlike traditional sales where ABC means "Always Be Closing", in Ultimate Selling the ABC stands for "Always Be Connecting".

The road to connecting begins with your very first communication, and it continues through successive interactions to the win/win proposal. And, connecting happens more deeply from that point on.

Let's talk a little about initial impressions, as these create an impression. These impressions cause prospects to make a quick judgement about you and the value you may bring to the plate. How you carry yourself, how you express your self confidence and the types of statements you make and the types of questions you ask validate their impressions and judgement.

Let's face reality: people want to do business with successful people, and sometimes you don't get the chance because you don't get past their first impression. So, if you want to connect, here are some of the key attributes that you want to demonstrate from the very beginning and throughout the entire process.

Key Attributes for Connecting

- Appearance: you've heard 'dress for success' said many times. Regardless of what anyone might say, it's impressive!
- Attitude: calm self-confidence expressed with an air of humility and empathy, not from an arrogant or narcissistic attitude.
- Competence: knowledge of the market, your product and service and awareness of the value you bring to them.
- Personality: be personable, but remember—it's a professional business relationship you are pursuing, not a social meeting
- Acknowledging every player: be cognizant of all the parties and acknowledge them as you come in contact with them, as you never know what role they might be called to play.

- Perspective: when communicating, making suggestions or asking questions, always do so from a point of view that's beneficial to them and providing alternatives whenever possible. Rarely will anyone reject anything determined to be in their best interest. So, think of it that way and phrase your wording in any communications accordingly.

Figure 10-1 Connect

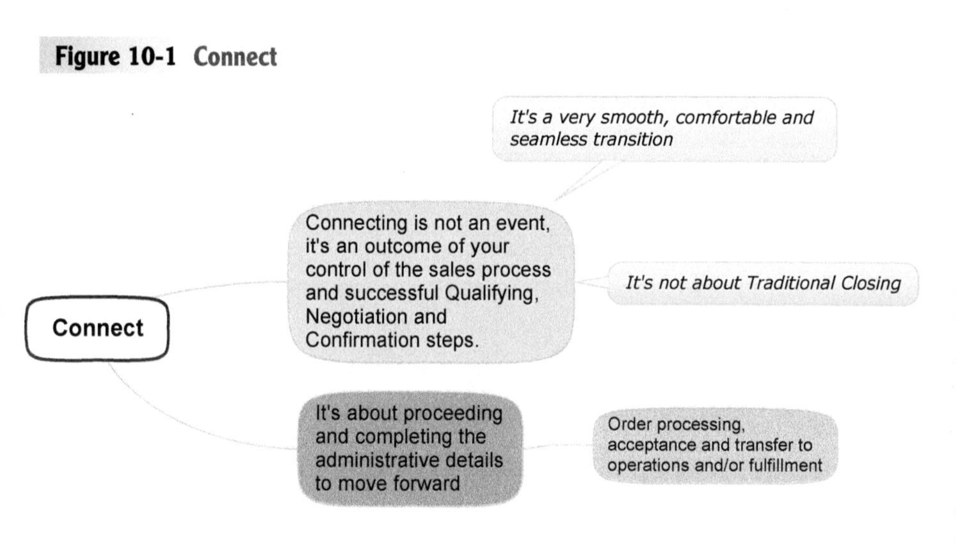

Ultimate Selling is tightly focused on what to do when you're in contact with prospective buyers and stakeholders, so it's critical to be fully prepared and ready to engage and connect throughout the entire sales process, and via any means of interaction. Anything short of full preparation will dilute your opportunity for success and eventual connection.

Each and every time you interface with the prospective customer you can build on your business connection with them. This should be subtle, and accomplished by the types of questions you raise, or your recommendations to the prospect that they view as important to them, ones that will enhance the process or their current situation. For example:

"Mr./Ms. Prospect, since you're just learning about our services, let's review your requirements together and we can figure out what should happen next."

This gives you the benefit of their involvement in the process and therefore, hearing their ideas and buy-in. Every time they say yes, you have created another building block for connecting. And they just agreed to do several things you want but—in their mind, it was their decision. Very subtle, but significant!

Ultimate Selling's approach is all focused on them, how to make their life easier by doing some work for them, providing additional expertise and insight and by NOT DOING a proposal at the wrong time. Trust me, if they object to this, then there are other issues, probably significant, that need to be clarified and addressed.

As you continue to qualify, always drilling down for a complete understanding of their needs and related impacts, you continue connecting in a big way. Be reminded to not rush this process. Listen intently and stay focused on them until you fully comprehend their issues, as this will be the bedrock foundation of your proposal.

When you finish Connecting, when it's time to shake hands, appreciate each other efforts and they have signed paperwork, congratulations. You really did connect. You should be feeling pretty good now, pretty confident and pretty inspired at this point.

You've reached a transition point in the book, at which we need you to decide what is best for yourself. In the next few chapters, we are going to delve more deeply into Preparation. Our intention is to set you up so that you have an incredible amount of control to leverage in the selling game. Up to this point, we've talked about how to control the sale from an activity (interaction) standpoint. For the next three chapters, we're going to talk about controlling the sale from a 'depth of knowledge' perspective.

Now, what is best for you at this point? Pause, practice what you've learned so far, and move into the expert territory some time in the future? Or press on, practicing what you've already learned, and learn how to map out solid, strong strategies designed to put you in maximum control and increase your selling success even more?

Remember that others will be reading this book as well. Some of them will continue through Preparation, Revisited. Who will have the advantage?

Whichever you choose, one thing is not optional. You've taken the time to read and understand Ultimate Selling concepts so far. You can easily see the value of how changing the game slightly is actually a huge change in your favor. So whether you have the discipline to head for 'expert' territory, or you need time to pause, the one thing you must do is keep practicing what you have learned. Come back to the first chapters in a little while, and read them again, then practice some more. This is a radical change and will take practice for you to be fluent using it. Ultimate Selling cannot be something you read in a book; it must become second nature for you. Practice to be your best.

Chapter **11**

PREPARATION, REVISITED

We hope that you've had a few "a-ha!" moments as we've shown how to control the selling process through simple conversations. You've seen the strategic strength of slowing interactions through deeper qualifying.

We also hope that the high-level structure of Qualifying and Negotiating on Interest, Qualifying on Money, Qualifying on Decision and finally Confirming and Connecting makes sense to you. P + Q + N + C really does equal C! You can practice these strategies in the simplest of conversations. The more you practice the more skillful you will become. As you begin we will simultaneously deepen the foundation of your sales skills by revisiting the topic of Preparation.

We discussed Preparation briefly earlier in the book. We feel strongly that good preparation is a huge factor in being ready to qualify and negotiate successfully. Now that you see the value of deep qualifying and negotiation, we trust you will be eager to learn even more about preparing for success. Before we get into specifics, let's review a little more about how and why preparing deeply makes sense.

Despite all good advice and common sense to the contrary, unfortunately, too many salespeople still don't prepare enough.

So if it's so important, why **do** so many in sales fall prey to not preparing well? Is it:

- Lack of self-confidence
- Lack of belief in their product and service
- Lack of belief in the opportunity
- Procrastination
- Fear of failure
- Afraid of the answer
- Poor work ethic
- Lack of discipline

- Think sales is supposed to be all fun, or easy
- Simply don't know how

While some or many of the above may have been true for you up until this point, I believe that time spent on foundation work can change everything for your confidence and posture.

What a Difference a Little Knowledge Can Make!

It is time to deepen the foundation of your skills.

The first half of Ultimate Selling was about *managing interactions*, helping you to slow down the 'rush-to-propose' in favor of becoming a trusted resource. We focused on the specifics of what takes place between you and a prospect, and how to change the game in your favor (and theirs as well.)

The second half of Ultimate Selling is about *managing information*. Besides intentional interactions between you and prospects, Ultimate Selling includes having a strong foundation that supports well-designed processes.

What do we mean by having a strong foundation? What do we mean by having good processes? How do you get them? And how are they related to Ultimate Selling?

A foundation, in building terms, means a stable, strong base on which the whole building stands. It's the same thing in Ultimate Selling. They differ only in that a building has a visible base, while in Ultimate Selling, your strong foundation comes from your knowledge and preparation.

Well-designed processes are like a well-run facility. Things are where they belong, easily accessed when needed. A sense of calm order prevails.

You've heard the famous quote *if you fail to plan, you plan to fail.* Failing to prepare really can be seen as preparing to fail. Preparing is about doing everything you can to control outcomes successfully rather than leaving things to chance. If you remember in Chapter Two, we talked about the importance of preparation. Thorough preparation builds confidence. It can best be seen by looking at the lack of preparation: you can almost always tell when someone is not prepared. They flounder, they make mistakes, they try too hard. You don't want to be that person…which means you do want to be the person who is fully prepared, unflappable, confident that whatever comes up, you will be able to handle it.

Maximizing control over the outcome doesn't mean things always happen our way...but it does mean we've done our best to ensure successful outcomes.

If you wanted to become a winner at poker, you would not just show up at the table, buy your chips and then leave it to chance, or would you?

You would want to know there are fifty-two cards in the deck with four suits, the value of each card, what were the rules of the game, and more. In poker, knowing strategy can mean things like knowing when to split a pair and when not to, knowing when to raise your bet. Knowing strategy, knowing the game rules, odds, and knowing your competition can maximize your opportunity for success—just like in poker.

Those familiar with country music may remember a famous song called "The Gambler". Love it or loathe it, the song has some significant points that apply to Ultimate Selling:

You got to know when to hold 'em, know when to fold 'em,
Know when to walk away and know when to run,
You never count your money when you're sitting at the table.
There'll be time enough for counting when the deal is done.

Now every gambler knows that the secret to surviving
Is knowing what to throw away and knowing what to keep.
Cause every hand's a winner and every hand's a loser,
And the best that you can hope for is to walk away with all the chips.

Okay, yes, we changed the last line. Just checking to see if you were paying attention! But really, that's what we want to do: teach you not to count your money while you're sitting at the table. Focus on your prospective client. Know what to pitch and what to keep. And yes, we intend to change things so that you do walk away, the winner.

The song goes on to talk about reading people's faces and knowing what their cards were by the way the held their eyes. Unfortunately, in a game of cards, you can't ask what their hand is. But as you've seen, we *can* ask questions like that when selling.

Whether we *choose* to do so or not, we *do have* the option of asking clarifying questions in order to validate not only what their eyes and faces are saying, but more important, what their dialog and actions mean.

So, the first (and maybe most challenging) part of preparation is what you've already learned: deciding to be strong and decisive about asking anything you need to know and ask.

Now, let's broaden the scope of your preparation to include organizing information that becomes your foundation for qualifying and negotiating.

Here's a breakdown of what we mean:

Table 11-1 Knowledge Creates Control of the Sale

Knowledge	Tools We Use To capture Info	Underwrites
Maintain accurate information about each prospect and status.	Prospect Profile	Decision Negotiation
Possess a deep understanding of the product or service you are representing. Know all strengths and weaknesses	SWOT	Negotiation
Know strengths and weaknesses of your company. Identify which are within your control and which aren't	SWOT	Negotiation
Know strengths and weaknesses of competitor offerings	SWOT	Negotiation
Know every step you typically need to sell your service or product	The Selling Process	Entire Process
Know your goal for each interaction and what resources and approach you need to accomplish it.	Intentional Interactions	Entire Process

Taking time on these tasks may sound daunting, but it's absolutely not. Much of it is done one time, refined periodically and used over and over. The other parts such as creating their prospect description or preparing for interactions are skills that, once you learn them, get more and more satisfying to use. You actually find yourself wanting to learn them better!

In the next few chapters, we're going to cover specific knowledge areas: knowing about your prospect, knowing about your product and company, knowing about your own selling process, and knowing how to intentionally plan interactions with prospects.

You can pause here in the book and just apply what you've already learned and come back to this later...or you can continue through the book now to delve even more deeply into controlling the sale from a knowledge perspective. The more you take the time to learn, the more successful your sales will be. Have fun with it and just explore! Let's get rolling.

TIPS

- Set aside dedicated time to work on each of these chapters and exercises
- Remove all distractions so you can focus on the task at hand and accomplish it better and faster
- It's okay if you don't finish each exercise at one setting, just make sure you schedule your next block of time and keep at it
- Some will be harder than others. But the work gets easier as you go along!
- Get the advice of others if you're stuck or have someone to look over it when you're done
- Start with rough drafts...and come back sometime later to rework the exercise!

Chapter 12

SWOT Analysis: Strengths, Weaknesses, Opportunities & Threats

If you're a large company, your first thought maybe ***"hey, why should I conduct a SWOT analysis? That's marketing job!" And, if you're part of a small company, you may have sighed "But nobody has the time!"***

For many of you, a SWOT of your products or services compared to competitors' may be readily available as a support tool created by marketing or a sales-support department. If you don't have access to an up-to-date SWOT, you need to create one. You most definitely need knowledge about how you stack up against competitors and the market conditions if you want to be in control of the selling process.

Why? Because your success and income depend on it.

A good SWOT and your thorough understanding of it allows you to be ready when you're qualifying or negotiating. Let's say you need to negotiate on price. The SWOT points out that the durability and reliability of your product actually reduces the overall cost. This can be delicate because they may not be concerned about overall cost, only today's costs.

Qualifying on interest? You've learned that your prospective customer is an advocate for the environment, human rights, or safety records. You know that your own production facilities have implemented significant controls to protect workers, or that your company implemented scrubbing towers to have a net-zero toxin emission into the air. These things can matter, as a client weighs costs and needs against ethics and emotional priorities.

Let's consider some questions.

Is the SWOT current? Is it regularly updated to reflect changes in competitor's products? The market? How is it updated? Or is it out of date or non-existent?

Does it incorporate input from real-life interactions with actual users? And you know there's a huge difference between someone behind a marketing desk conducting an academic exercise and being in the trenches. How confident are you with the information?

If you're going to improve you're sales successes, you have to first understand why some AREN'T.

If management needs to address competitive issues you can't expect them to react to your verbal input (sometimes perceived as excuses). You need to build a legitimate business case as to why you're winning less. A good SWOT can clearly draw the line between a sale lost due to competition that served the prospect better, and a sale lost due to ineffective effort by the salespeople or problems with the company backing them.

A good SWOT about your company, besides being just about your product or service, can serve you by addressing weaknesses and winning more sales. It can help your company, stockholders, and even customers, by making your company stronger, more profitable, and sustainable.

How readily does your company incorporate input from the "street" about user experience and competitor values? If you're living on your sales earnings, it makes sense for you to provide accurate competitive intelligence to yourself, your company or organization—and to appreciate having it!

Remember buyers have quick access to information not only about products and services, but also about other corporate strengths and weaknesses. Customer reviews will rate your product or services favorably or unfavorably. Unless you're a private entity, corporate financial information is a few mouse clicks away. Your company needs to be accountable to you as well as prospective and existing customers, not to mention investors or stockholders, based on real-world responses.

Let's look at how a strong SWOT can defend your product. Say you sell a software product that must be installed on a server locally, while your competitors offer web-based versions.

Sounds as if they have the advantage, doesn't it? But if you have done your homework, you can offer clients some of the hidden, *not-touted drawbacks* of a web version: while it's much cheaper up front, the long-term costs can be far, far, higher. Explore with your clients if the web versions offer strong reporting options...or are reports limited and difficult to customize? What if you want to switch to a different product...can you get all your data from the web version? In a bad economy, remember that you never "own" the application...so if you fail to make regular payments based on revenue or cash shortfall, you could lose access to your database! Not so on a server version; you can opt to let the software get out of date, but it's still usable.

This example came from a real life SWOT carefully compiled by a salesperson (not from the company's SWOT) and proved a huge factor in negotiating with prospects and a successful outcome.

The message is clear: the more you know, the better you can respond intelligently to competitor attacks or prospect concerns.

Now, on the other hand, you may have some legitimate frustrations with your company or its product or service offerings. A good SWOT may be the catalyst for your company addressing real issues. It's okay to be honest about them—but remember that you are being paid to sell their product or service and are a representative of the company! In these cases, you have a responsibility to sell what you have based on it ***meeting your customer needs and requirements***. If you can't accomplish this, then they *should* choose a different solution. You'll have no need to go to Confirmation.

Having the ability to let your customers know the important questions they should ask (of you and of your competitors) helps you solidify being the trusted provider. It's important to help your prospect learn the questions and not just spoon-feed them the answers. Telling them "our product is better because it only needs to be maintained every two years!" is not nearly as strong as having them reach that conclusion on their own with your guidance: "What's the effect of maintenance down-time on your business?

A good SWOT allows you to direct the conversation to where you know your strengths are. For example, have you reviewed the screen refresh times/rates and what is your minimum requirement? "Oh, A-Z Software hasn't gone over screen response times? Are you aware that slow screen responses are the number one complaint of almost every web-based software? It's partly your "last mile", partly their servers, and partly traffic based. There's no easy solution. My question is, what's the effect of having your staff waiting for screens to change?" "What's acceptable to your company?"

You can see that you've directed your competition to a possible disadvantage in your prospect's eyes.

A good, current SWOT also allows you to identify real obstacles to your prospect. "We will have a delay in delivery schedules in March because we're upgrading our manufacturing facility." Issues identified clearly and unemotionally with them allow you to plan together. Again, an honest assessment of your company/products SWOT and your willingness to share openly with them greatly increases their trust in you and their desire to work with you.

You can always lose a sale because of non-logical reasons such as politics or favoritism. Except for those situations, you can rest assured that prospective customers prefer to buy from those they trust and believe will deliver.

The most important thing is your prospect's ability to be aware of all the pros and cons so they can make an informed and intelligent decision.

Hopefully, I've motivated you to do the necessary work to complete a SWOT analysis.

Please note: if you'd rather simply practice Ultimate Selling from a dialog standpoint, you can of course opt-out of this exercise. Just be aware the these recommendations come from years of experience in selling. The more of the Ultimate Selling techniques you master, the greater your earning and job satisfaction potential. So, please make a commitment to work on this in the future.

Background and Overview of an Ultimate Selling SWOT

Our goal is to have you create a SWOT analysis that is not focused solely on products or services sold, but on sales-ability (a differentiator) and fulfillment. How capable is your product, your service, *...and your company that supports them*?

The Four Key Components of a SWOT Analysis

Strengths, Weaknesses, Opportunities and Threats divide by four factors. They are things seen as generally positive, or generally negative. Likewise, they are either something you or your company can control and change (maybe with some effort!) or something beyond your control.

	Positives:	**Negatives:**
Controllable:	Strengths	Weaknesses
Not Controllable:	Opportunities	Threats

We'd like you to draw a quick grid on a blank piece of paper or writing tablet as shown above, leaving lots of room to write under the words Strengths, Weaknesses, Opportunities and Threats.

Strengths

What are the positives in your organization over which you have some control? What things do you personally or the company as a whole do well? Strengths include things like your good reputation, tangible assets, credit terms, control of raw materials, plenty of human resources available for production, skilled resources in sales, customer support, 24/7 helpdesk, technical and engineering personnel, sustainability, good safety and environmental records, copyrights, patents or any valuable resource within the business. In essence, you're looking to define any and all of the strengths that add value to your company.

What are the positives in your product or service, that, likewise, you or your company can control? What do they do well? List them all.

Weaknesses

What factors *detract* from your ability to be competitive and are *within you/your company's control*? They may include lack of expertise, rapid staff turnover, poor design review processes, cash flow problems, service and support issues. You gain tremendously by identifying your corporate weaknesses and correcting them so that you can compete better.

What factors that detract from your product or service are within your control? High cost (assuming you have flexible margins) could be one. What might be others? Please make a list of these in your grid.

Opportunities

These are external factors beyond your control that still affect your company in a positive way. Is the demand for whatever you provide increasing? Do you provide heirloom seeds to organic growers...and more people are demanding organic products? These are external market conditions that create opportunities that your company can fulfill. Please write a list for your company.

Threats

Threats include factors that are also beyond your control but which affect your company in a negative way. Threats are things like patent expiration timelines, government regulation, economic downturns, consumer behaviors, cost controls, suppliers price increases, becoming-obsolete services.

A few examples are wristwatch and clock manufacturers, who are affected by cell phones becoming the timepiece of choice. The Gulf oil spill affected many, many industries: fishing and shrimping, realtors trying to sell properties along the Gulf coast, home improvement, decorators, banks, mortgage companies, the vacation industry and all who provided goods and services to these companies.

What threats are you aware of that are beyond your control, but could affect your ability or your company's ability to create and sell products and services? Please jot them down.

Let's play a little to give you a break.

If you were standing outside the doors of your company and there was a big billboard listing strengths and weaknesses, opportunities and threats, it might read like this:

Strengths	Weaknesses
We have no corporate debt! We have beautiful showrooms and meeting facilities	We can't hire enough people to manufacture widgets on the night shift because we don't pay enough!
Opportunities	**Threats**
We have a growing demand for our widgets because every new home is required to have a sprinkler system!	Folks in China are starting to manufacture widgets at a cheaper rate!

Standing outside your company and looking at this billboard, you'd feel enlightened. It may even help you decide if you wanted to continue working for this company or not.

Likewise, your clients may appreciate knowing the same kinds of things. They're not looking to be employees...no...but they ARE looking to have a good relationship. Would they rather you tell them upfront about the production problem, or find it out later when they've got 400 homes that can't pass inspection for sprinkler widgets because your production is too slow?

All companies face issues and challenges. It's how they react that's the difference. Being open allows you and them to work honestly together.

Once identified, you can assess and rank how much each item, especially the negatives, affects your ability to sell effectively, and most importantly, you can decide how and when to use this information in the qualifying and negotiating steps with your prospects.

With practice, you'll learn when and how to share information about your strengths as you qualify prospects (so they want you), and about your weaknesses in negotiation (so there are no unpleasant surprises), to position yourself as an honest, trusted advisor early and throughout the selling process.

When appropriate, you can address weaknesses either internally with your own company (raise nighttime pay rates) or externally with clients (address Chinese costs versus your quality and safety-design patents) in a proactive, not reactionary manner.

Summary:

It's important to analyze your internal strengths and weaknesses as compared to the external opportunities and threats. With insight, you can best take advantage of opportunities, and minimize any threats that could reduce your sales.

Documenting Your SWOT

For your reference, we have included some examples of a SWOT Analysis, represented by different formats for capturing the data. In **Figure 12-1**, we use Mind Mapping. In **Figure 12-2** we use a Graphic Quadrant and in **Figure 12-3** use a spreedsheet.

What's important is to incorporate all the applicable information of your internal strengths and weaknesses and the external opportunities and threats. It's not important that you elaborate on each topic, but simply create bullet points upon which you can expand later.

Time to schedule your exercise assignment

Schedule a block of time with yourself to jot down your thoughts about your company's strengths, weaknesses, opportunities and threats. As you're looking at them, even just a scribbled version, you'll immediately see how the SWOT knowledge can positively affect sales.

A final thought about SWOT: *If you've given your company management accurate, intelligence and specific suggestions based on the SWOT, and they fail to act on it in a manner and timing appropriate for the continued success of the company, an Ultimate Seller should experience real concern! Is your company being as accountable to you and your customers as you are?*

Figure 12-1 Mind Mapping

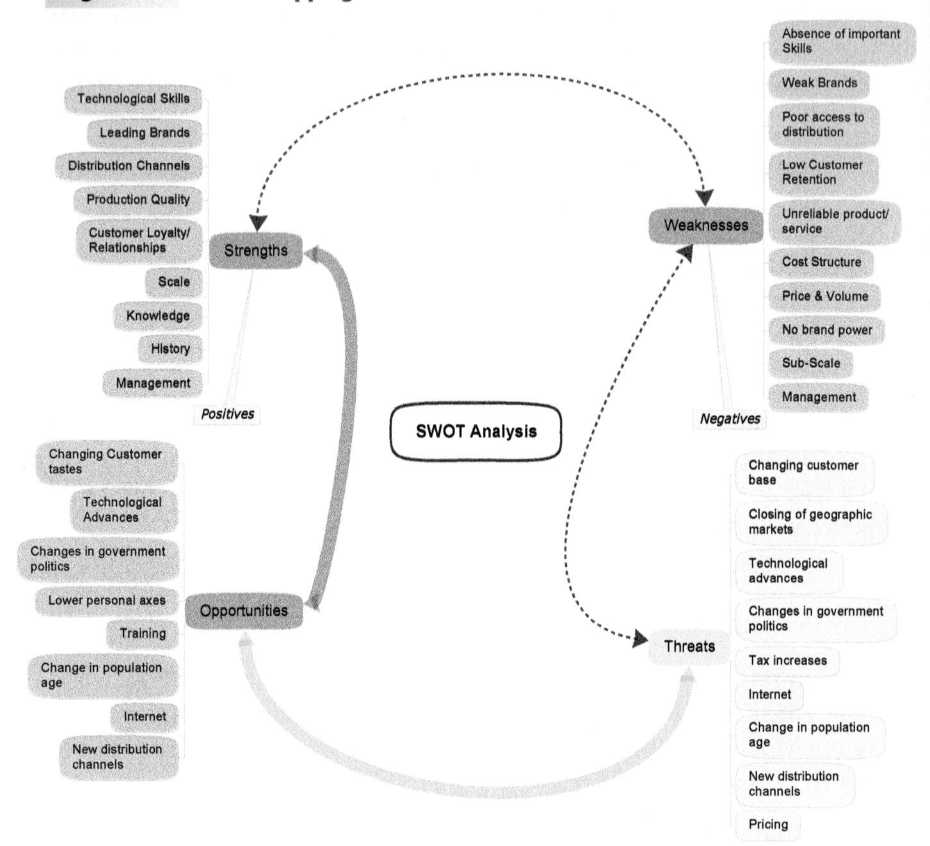

Figure 12-2 Graphic Quadrant

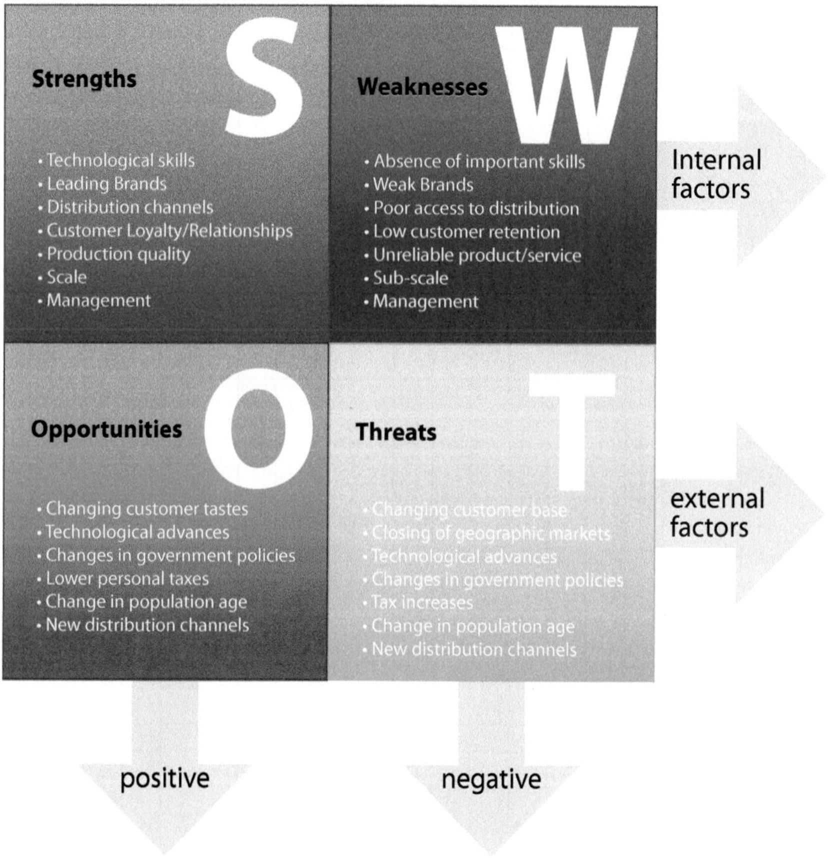

Strengths S

- Technological skills
- Leading Brands
- Distribution channels
- Customer Loyalty/Relationships
- Production quality
- Scale
- Management

Weaknesses W

- Absence of important skills
- Weak Brands
- Poor access to distribution
- Low customer retention
- Unreliable product/service
- Sub-scale
- Management

Internal factors

Opportunities O

- Changing customer tastes
- Technological advances
- Changes in government policies
- Lower personal taxes
- Change in population age
- New distribution channels

Threats T

- Changing customer base
- Closing of geographic markets
- Technological advances
- Changes in government policies
- Tax increases
- Change in population age
- New distribution channels

external factors

positive

negative

Source: www.gliffy.com

Figure 12-3 Spreadsheet

Category/Topic	Strength	Weakness	Opportunity	Threat	Notes:
Design			X		
Engineering and/or Configuration		X			
Product, Quality	X				
Product Configurability				X	
Project Management			X		We provide more on-site PM than alternatives
Support, On -site	X				
Training		X			
Product, Life Cycle			X		
Financial	X				Excellent financing options. Self or 3rd party
Agreement T&C'S					
Employee skill levels					
Management					
R&D Innovation					

Step 1 - Insert the category or topic

Step 2 - Check all boxes that apply

Step 3 - Insert applicable notes

Ok, we've looked at how we sell in the selling process.

You've looked at your company's SWOT.

Now: How can the selling process be pruned, beefed up, modified to help us close more sales?

Can your SWOT be used to help your company close more sales? What needs to be done and managed? **Let's start the first small steps of fitting some of these pieces of information together.** In our next chapters, we will look in depth at our selling process.

Chapter 13

Owning Your Selling Process

As you worked on your SWOT, you considered the strengths of your company. One strength may have been having good, well-managed internal processes.

A process that some companies think about and some don't is their selling process. Having a clear selling process is a huge advantage. We'd like to go over our recommendations for you.

First: What Is a Selling Process?

You may be thinking, *we don't have a formal sales process,* or *we don't have a complete one.* Whatever you are thinking, you *do* have some kind of sales process you follow.

If you were teaching a new employee the selling process in your business—how to do their job! what would help describe it to her or him?

You might pull together a list of actions: do a site survey. get a users list, decide who needs training. Send samples. Get permits. Take documentation photos. Check zoning. Anything that's necessary for your sale.

It's not your "style" in getting something done, it's the actual steps required to make a sale, from the initial interaction to agreements to be signed, who must sign them, the tools needed to achieve certain steps and objectives, the demos, the samples shipped, approvals, etc.

That's what 'the selling process' is. In your business, the information and actions required to get a sale finalized. It's in your head. We'd like you to write it down.

First, why does it matter that your selling process is written down?

A well-defined selling process offers:

- A starting place for having things under control, being calm, and presenting an extremely professional and trust-inspiring appearance to your prospect.
- A roadmap to successful sales
- Good organization helps keep you in the driver's seat rather than your prospect or your competition
- Faster ramp-up for new salespeople and sales support staff
- Common understanding of what it takes to sell your product or service, so that your company and resources can work together more effectively
- A quick list of the necessary tools and resources needed and available
- The ability to monitor the progress of the sales opportunity and the ability to monitor and modify your process whenever it's flawed

Of course there are some variations for each sale—but as long as you know what steps HAVE to occur in order to reach success, you are much more likely to react appropriately to whatever is thrown your way.

Please take out the notes we asked you to jot down in Chapter 3. Look at the steps you listed that it typically takes to close a sale. If you think about your business across a year, and listed the elements of every sale you'd worked, you'd find many of the same elements, the same steps, the same needs, repeated again and again. They are:

- **WHAT ACTIONS** need to be done to make it happen
- **WHY** each needs to be done
- **WHO** needs to do it
- **WHEN, WHERE , HOW** they generally happen

Selling contractual welding services to shipyards may require more steps than selling composted mulch to garden centers. You'll know from doing your business what's needed. Again, think of it from the viewpoint of what a brand-new-employee requires for success and just fill in the blanks for the elements above.

That's how you get started documenting the selling process for your business. Tedious? Maybe. Valuable? Very.

You'll create something of great value for yourself and your company: a clearly-documented selling process that can be used and improved.

Understand the concept? Okay, don't do anything yet. First, get your calendar and schedule time to complete the exercise. If you can make time now or if it's not until next week or takes a few sessions, that's fine. Just make sure that you allocate time, and make sure it's set in stone. Make sure it's *not* in prime selling time. This is a real appointment, with yourself.

As stated earlier, we encourage you to complete all chapters in sequence—but this is your journey. Work on this exercise when you choose.

Pen to Paper Time

Let's have some fun doing this exercise! We'll get your input, and then use what you already know—but with a huge twist: you are going to use these steps differently than you ever have before. The selling process document that you create will be part of something we've alluded to in Preparation Revisited, Chapter 11. It's not a passive document. It's part of the foundation *to put you more and more in control of the selling process*. You will be able to leverage your time more effectively by combining the information about your selling process with our structure for qualifying and negotiating.

Let's start by expanding the list you jotted down in Chapter 3 by inserting all your sales processes into the rows of the form you will now create. It will be five columns across:

Activities (WHAT ACTIONS need to be done to make it happen)

Objectives (WHY each needs to be done)

People—Yours (People who are inside or in support of your company)

People—Theirs (People at the prospect's organization or connected to them in some way)

Mechanism and Tools (WHEN, WHERE, HOW it generally happens)

Don't think about it. Just put your pen or pencil on paper, and replicate the form below. Or, create one in Excel or a Word document, and begin the process of writing it down.

Note: We invite you to review in some detail the completed **Sales Process Example** in the end of this chapter. It is recommended that you refer to the sample as you explore the various elements and *DO NOT* be overwhelmed with the amount of detailed information.

As you develop your own document, you may add or subtract based on whatever elements you consider essential for successful sales of your product or service.

Don't take shortcuts as this is your road map. It's your personal 'yellow brick road' to sales success. More important, if you get lost you will be able to quickly get back on track. Let's begin with the explanation of each Column.

Table 13-1 Your Selling Process

Activity (What)	Objectives	People, Yours"	People, Theirs	Mechanism & Tools "How

Column One: Activities

Make a list of all activities that can and should take place from the beginning to the end of your selling process.

This is your rough copy, but as you go, try to organize your list of events and activities in sequence. Order them so that they show a progression, things that you typically do in sequence as they advance from a suspect, to prospect to qualified prospect, and ultimately, a new client or a new sale to an existing customer, (and on to installation if that's part of your responsibility.)

If you're selling land to developers, you may write things like "check about zoning, development rights," "get a big sign on property"… "get perc tests"… plus a title search "updating mortgage rates" "credit check", etc.

If you're selling jet engine designs to Boeing, your list might include many, many steps to have them review technical designs, conduct workshops with end users, participate in bid conferences and processes, price components that you will outsource, show test data, etc.

If you're selling tree trimming services to shopping malls, your list may be pretty short: "do an onsite quote"… "check land for equipment access"

... "verify ability to pay" (and that could be just a visual of the people and property you visit and confirmation on terms) "make sure you identify equipment needed" (such as a lift with a retractable basket, or a crane)

Remember, we know that your list will be longer or shorter generally depending on the price tag of the products or services you sell. Big price tag? Usually a big list and longer timeframe: the more steps or sub-steps are needed to confirm for your prospect that your product or service is what they need. Smaller price tag? Smaller list!

Don't make your list overly complicated, but DON'T leave things out. Better that your list is longer rather than shorter...because it means you are really thinking about it. You can always trim it down later.

Do you always need to do a credit check, or line up leasing or financing? Is there a board of directors or system users who need to see a demo or approve the project? Do you have swatches to approve? Environmental concerns? The list can be long, but you know what they are. Again, think brand new salesperson training. What would you list if you were teaching them?

Here's a twist: what would you want to see and know if you were the buyer? What's important to them, even if not to you?

Brainstorm and write the activities down as they come—and be absolutely sure to write down activities that you typically know you *should* do, but don't!

Column Two: Objectives

In the next column, we want you to write down the objective for each activity listed in column one.

Oh dear. We can hear you thinking again. "This is too basic. What a waste of time!"

Yes, it's incredibly basic and no, it's not a waste of time.

We're going to say it one final time here: if you're closing more than 70 or 80% percent of your needed sales, you may want to close the book. If not, you're here for a reason: you want to make more money, and you want to feel better about your work, win more and actually enjoy it.

What do you have to lose if you work through the exercise *the right way*? What do you stand to gain?

Your objectives for each activity must be very straightforward. They must say specifically what the underlying purpose is for the activity, a purpose which must be achieved to advance to the next step. While the Activity says specifically *what* you and/or the prospect or suspect must do, the objective states the reason for it.

What *is* the point of each step? There's a reason you do it. You show samples to create a desire for one of them—and you can learn what they liked and didn't like. Sound like qualifying for interest, maybe?

You take a CEO to lunch…to build rapport and establish credibility with them. You may provide references or testimonials…again, to deepen your credibility as a viable source they can trust. Whatever the activity, you're doing it for a reason. Write that down. That's your activity's objective. Again, don't get hung up on if it fits perfectly. We'll clean up later. This is just practice.

Now, ***get this idea cemented in your head*** because as a tenet of Ultimate Selling, ***once you know the objective for an activity, the right to proceed to the next step must be earned by you and your prospect by fulfilling the objective completely.***

It's not automatic, and you don't get to skip steps to rush the selling process.

Let's talk about that for a moment. Rushing to money—skipping steps, or not accomplishing the objective of an activity—is an all-too-frequent trap that average salespeople fall into, in the hopes that somehow they'll get the sale if they can make it happen *faster*. You know in your head and heart that it does not work. But then again, maybe you're an eternal optimist: fantasy generally beats reality! Maybe you like to focus on slim hope rather than real results.

If you're fortunate enough to pay your mortgage or rent, utilities, car payment, health care, groceries, education, recreation and savings with 'hope cards', stay your course. On the other hand, if you want to be proactive and assert some control over the outcomes and have cash in lieu of hope cards, then I suggest it's in your best interest to do this preparation work. It's your call. It's your bank account. It's your security and your future. Either way you have to take personal accountability for the outcomes.

Lecture over. You're still with us. Good for you.

Now that you've listed the steps needed and the objective for each step, you're at the easy parts. The hard work on this one is done! All that's left to do is identify the Who (internal and external personnel) and the How (Mechanisms and Tools).

Column Three: People (Yours)

Whose help in your company do you need for the sale?

Who outside your company—a manufacture representative—a third party—is important?

Identify all personnel available at your disposal—either in your own company or in partnership with your company—that will be resources to provide help.

Column Four: People (Theirs)

Who are the people at their company who need to get on board? Which of their business associates are somehow part of making the event happen successfully? They may be friendly personnel or unfriendly: they may be friendly to your competitor, they may be uninterested in spending money, or they may be helping you.

Never discount *any of the non-decision makers or* external personnel. It's possible for the boss to call anyone—a trusted administrative assistant, a bright intern, a company founder, a business friend, a consultant, or their favorite palm-reader to ask their opinion on what you're presenting. After all, everyone wants reassurance that they are making good decisions—and if the salesperson can't fulfill that role, then someone else will.

Note: part of the big picture of this book is to help you become the one who fills that role as you progress through your selling process!

As you identify people who are important to the sale we have a small, but important recommendation for you. Whether they are directly involved in the sale or decision making process, i.e. receptionists, executive assistants or other employees or related parties, their bank officer or whomever, *acknowledge people regularly as you're working the opportunity.*

Shake their hand. Let them know you are aware of them. Many times, if you acknowledge someone, that in itself is all that's required to be remembered and if their input is requested, you will get their support—and it's icing on the cake if you can take it beyond that.

That concludes the 'WHO' parts.

Column Five: Mechanisms & Tools

This is the nuts and bolts of how things typically happen—where it takes place, what physical items are needed, etc. List it thoroughly, just as if you were teaching a new team member so that you do not short change yourself in the process

Now that you've got a rough draft, find peers or a manager to review your sheet, to look quickly over your listed activities, your objectives for each, who the players are—and have them give you their input and advice.

Done? You've completed the draft of your selling process? Congratulations!

Now look at your list. I'm betting that the longer you wrote, the more the list, especially the objectives, got to be about what's needed from the prospect's point of view. The more you really think, the more you're like an Ultimate Seller.

The next step is to take the selling process, integrate it with your SWOT, and use it to plan intentionally to hit every one of those listed objectives with your prospects in the chapter on Interaction Planning.

Once again, this may feel like it's very time consuming. We trust that if you're selling office supplies and equipment on commission, you'll know that the whole process may be very quick compared to selling beach erosion control equipment and services to a shoreline city, where county and state governments are involved! But when you take the time to focus and identify the required steps, large sale or small, you should see your win ratio (and your bank account!) go up in proportion.

Sales Process Example (Example is a Micro - _You need to scale yours up or down as required by your product or service_)

Note: ***The becomes your roadmap for successful sales and should be referred to throughout all steps of the process***

ACTIVITY (What)	OBJECTIVES (Why)	PEOPLE, YOURS	PEOPLE, THEIRS	MECHANISMS & TOOLS (How)
√**Build a Sales Pipeline**	√To generate a "Suspect" opportunity √To screen using you ideal customer filter √To ID/pre-select those that warrant being a qualified suspect	√Salesperson √Marketing Information √Telemarketers	√Receptionists √Assistants √Other company personnel √Department Heads	√Lead Generation Programs - √Company Database √Client Referrals √Sales Lead, Prospected/Cold Call √3rd Party Referrals √RFP/RFI/RFQ √Seminars, Trade Shows √National Account List, etc.
√**Prepare via Research**	√To gather additional information to support a sales pursuit decision and used throughout the sales process √Verify information to date	√Salesperson √Telemarketer √Inside pre-screener √Sales/Marketing assistant	√Receptionists √Assistants √Other company personnel √Department Heads	√WEB site √Goggle! (Search Engines on company or individuals) √Telephone survey questionnaires √Telephone call using dialogue scripts √ ...***what do you typically use?***
√**1st Interaction**	√Schedule a face-to-face meeting √Schedule a phone appointment as required √Verification of current prospect information √Acquire decision-maker information √Get decision-makers to participate in 1st meeting	√Salesperson *Note: If you're starting at the top, do you entrust this activity to someone else or do you do it?*	The person/s at their company who initiated the contact, or if you are initiating contact: Executive Management, Executive Admin *Note: Start at the highest level you can reach*	√Prospects Description √Telephone Call √Appointment confirmation email √Formal letter √Templates with Special attachments, instructions or agenda to establish mutual expectations when you meet

Sales Process Example (Example is a Micro - _You need to scale yours up or down as required by your product or service_)

Note: ***The becomes your roadmap for successful sales and should be referred to throughout all steps of the process***

ACTIVITY (What)	OBJECTIVES (Why)	PEOPLE, YOURS	PEOPLE, THEIRS	MECHANISMS & TOOLS (How)
√Subsequent Interactions *"Looking for the solution without first understanding the problem is like working in the dark"*	**<u>Traditional Sales Reasons</u>** ***<u>(the reason for interactions is to sell ourselves)</u>*:** √Bonding & Rapport √To present our story and sell the benefits of doing business with us. √To introduce our company mission and philosophy. √To sell benefits of our product and service **<u>Ultimate Selling Reasons</u>** *<u>(the reason for interactions is to learn more about them)</u>*: √Establish self as a credible/trusted/leading resource and provider. √Differentiate self from competition √Schedule Needs Analysis √Schedule Executive Interviews √Define financial criteria, budget and funding √Define the decision making process. √ID, Recommenders, Influencers & Decision Makers √Identify current situation. √Identify prospect's goals and objectives. √Obtain commitment to act on efforts √Determine when product or service must be deployed √Determine if a Trial/Pilot Program required √Establish preferred communication modes, Phone, emails, best times *Note: Multiple calls may be required to achieve all the objectives. However, continuing the process without clear understanding is a high-risk investment of time & resources. Staying on track will produce the desired outcome with minimal if any, back end recovery mode*	√Salesperson √Application specialists or sales engineering support √Manufacturer Reps √Subject matter experts √Who else? *Note: Only involve persons essential to qualifying and advancing the sales process.*	√Executives √Department Heads √Board members? √Consultants? √Subject matter experts? √Other company personnel? √Who else? *Note: Target all decision maker/s, influencers and recommenders .* *TIP: Board Members or Consultants are determined by scale and/or who is required. (i.e. they have consultant or Board member brought you into the company)*	**<u>Ultimate Selling Minimal Tools:</u>** √Prospects Description √Meeting Agenda √Writing tablet or laptop √Meeting follow-up email confirming meeting results, next steps and commitments **<u>Typical list used in Traditional Sales</u>** √Corporate Overview (Powerpoint) √Offering Overview (Powerpoint) √Capabilities Presentation (Powerpoint) √White Papers √Demo CD/DVD √First Call Presentation √Quote Information Request √Company Brochure √Needs Analysis Form √Competitive Matrix √Other documents specific to your service or offering √Reprints √References & Testimonials √Case Studies *Note: In Ultimate Selling, we DO NOT use any of these without a specific objective to fulfill or validate a prospects need. And, your conviction it will advance the sale in a positive direction, or disqualify them from moving forward in the sales process* *REMEMBER: Its not about you, its about them. Learn everything you can about their needs & save it for the Confirmation step, when it can have the most impact, if they qualify.*

ACTIVITY (What)	OBJECTIVES (Why)	PEOPLE, YOURS	PEOPLE, THEIRS	MECHANISMS & TOOLS (How)
Note: Or any other activity or function required by your selling process that will advance the sale in your favor. Some examples are: ✓**Application Survey** ✓**Site Survey** ✓**Advanced Needs Analysis** ✓**Preliminary Proposal Workshop**	✓Identify unusual requirements for consideration and pricing. ✓Create team mentality, trusted advisor (Mitigate buyer/seller mode) ✓Solicit prospect participation in sales process and design of your proposal ✓Validate current situation and clarify the prospect's goals and objectives. ✓Gather additional information to be used in your proposal & recommendation. ✓Identify applications that are unique to your offerings. ✓Negotiate 'Options' either in or out of their proposal and Confirmation step. ✓Validate the "Players" and ID others who will be in the decision making process. ✓Introduction and buy-in from other participants ✓Position as "default" provider of choice. ✓Review present system, practices and results of the needs analysis. ✓Present solutions and recommendations for discussion and acquire prospects' concurrence ✓Secure Prospect Agreement on requirements. ✓Eliminate potential obstacles for a successful outcome and seamless transition to procurement. ✓Identify and secure prospect agreement on benefits, problems solved and financial impacts. How is it a WIN/WIN ✓Review decision criteria. ✓Financial analysis. ✓Address any client site visits or references prior to preparation of the final proposal.	✓List anyone in your company or sales team member (i.e. Factory Rep) who needs to be involved to successfully fulfill any of these next steps in the sales process	✓Any of the previously identified person or others who are required to fulfill the activity step so the process can continue. (i.e. subject matter experts) ***Note: If you proceed with steps in the sales process without the required resources and stakeholders, you will delay and diminish your probability for success.*** *TIP. Reschedule the meeting and demonstrate the importance of their participation. Thats Control!*	✓Prospects Description ✓Application Questionnaire ✓Pricing & Proposal Request Form or worksheet ✓Specific sales or technical documentation as may be required ✓Meeting follow-up letter confirming outcomes and commitments ✓eMail meeting follow-up template ✓Gantt charts ✓Client Site visits ✓On site demos ✓WEB demos ✓Manufacturer plant visits ✓Other proprietary tools at your disposal

Sales Process Example (Example is a Micro - <u>You need to scale yours up or down as required by your product or service</u>)

Note: ***The becomes your roadmap for successful sales and should be referred to throughout all steps of the process***

ACTIVITY (What)	OBJECTIVES (Why)	PEOPLE, YOURS	PEOPLE, THEIRS	MECHANISMS & TOOLS (How)
√ ***Demonstration of Product or Service*** *OPTIONAL?* *TIP. If you conduct an unnecessary demo, you can potentially undue the sale.*	√ Demonstrate that your product or service can fulfill their specific needs. √ Validate your stated facts throughout the sales process **Note: Unnecessary if you can move from the your proposal to an agreement of sale without a demo.** **Note: Unfortunately, many demos are scheduled by salesperson to fulfill their needs and are not necessarily required by the buyer.**	√ Anyone in your company or team member (Factory Rep) who needs to be involved to successfully fulfill any of these next steps in the sales process	√ All stakeholders, who will affect the buying decision. Be they recommenders, influencers and critical, decision makers	√ Prospects Description √ Client Sites Demonstration √ WEB demo *Note: Only demonstrate the features or capabilities essential to meeting and exceeding the needs expressed by the prospective customer. Anything more can potentially create obstacles or objections to concluding the sale.*
√ **Create the final draft of your proposal or value statement to be submitted to a Qualified Prospect**	√ To craft a proposal that incorporates all the critical elements from the prospects' perspective. √ All reviews and required approvals for pricing, Engineering, Legal, Financial, etc. √ Acceptance and approvals by any 3rd parties for special conditions. (I.E. Discounts or delivery dates from VAR, Manufacturer or others that may apply. √ Register and validate proposal as required by company √ Prepare documents or agreements to be executed	√ List anyone in your company or sales team member (i.e. Factory Rep) who needs to be involved to successfully fulfill any of these next steps in the sales process	*Note: This is not shared with their team until the Confirmation step*	√ Prospects Description √ Standard Pricing or spreadsheets √ Configuration worksheets √ Special Pricing docs. √ Summary √ Gantt Charts √ Executive Overview √ Warranties & Guarantees √ Post Sale Pricing information (I.E. Adds, deletes, labor rates, etc.) √ Customer References & Testimonials √ Payment Schedules √ Comfort Letters
√ **Schedule Final Proposal Presentation Meeting**	√ Bring the selling process to a close √ Gather all documentation in final form to use in Confirmation step	√ List anyone in your company or sales team member (i.e. Factory Rep) who needs to be involved to successfully fulfill any of these next steps in the sales process	√ All stakeholders, who will affect the buying decision. Be they recommenders, influencers and critical, decision makers	√ Prospects Description √ Proposal or Value Statement √ Agreements for Execution √ Maintenance/Service Agreement √ Software License agreements √ Agreement Schedules √ White Paper – Executive Overview √ Proposal Cover Letter

ACTIVITY (What)	OBJECTIVES (Why)	PEOPLE, YOURS	PEOPLE, THEIRS	MECHANISMS & TOOLS (How)
√**Proposal** **In Ultimate Selling, its time to CONFIRM.** ***Note: Sometimes you can conduct a Preliminary Proposal Workshop by presenting as such if this would help uncover any hidden agendas, etc. and then come back with your final proposal for Confirmation*****	√Review of their stated needs and requirements and affirm no changes √Affirm delivery requirements are unchanged √Present how your value proposition will meet and exceed this requirements √Hi-Level review of System, Service, Investment as required √Financial Presentation (if applicable). √Address any last questions that may surface √ASK IF THEY ARE COMFORTABLE WITH YOU SOLUTION/S - A YES IS TO CONNECT. √Thank them for engaging you and your firm √Ask how they want to expedite the agreements, PO, Money, etc. Whatever is required √Schedule next meeting to complete documentation requirements, **if needed.** √Schedule a meeting to transition from sales to the Project Management team (Installation, Operations, Service and support) for fulfillment with their appropriate stakeholders	√Any of the previously listed persons as orchestrated by the primary salesperson	√All stakeholders, who will affect the buying decision. Be they recommenders, influencers and critical, decision makers	√Conference facilities as required √Easel, white board, projector as required √Adequate copies of the proposal √Product literature as required √Customer References & Testimonials √Required Agreements of Sale √Agent Authorization templates for use on their letterhead (as required for authorization to act on their behalf) √Letter of Agency for execution √Non-Disclosure agreements √Letter of Intent, if required
√**Wrap-up meeting, if required**	√Address any and all incomplete items √Collect any documents, deposits, etc. *Note: Typically not required but an insurance step for undone or missed items*	√Any of the previously listed persons as orchestrated by the primary salesperson	√All required for completion of agreements, funds, etc	√All documentation missing or incomplete and not yet submitted. (I.E. Purchase Order, Tax Exempt Docs, Other)
√***Sales and Order Proceessing***	√Complete processing of the new sale. ***YOU CONNECTED***	√Primary Salesperson and others as needed. (i..e Contracts Admin, Finance)	√As may be required	√Order processing requiremnst & procedures
√**Transfer from Sales to Operations**	√Hand off from Sales to Operations or your fulfillment team	√Any persons required to complete objective as orchestrated by salesperson	√All internal personnel. (i.e Project Mgr. etc.) √Person who has authority to make decision during the fulfillment stage.	√Copies of all agreements that reflect the deliverables and Terms and Conditions of the Sale

Chapter **14**

The Prospect Description: Critical Information At A Glance

You've got your selling process written out, and are learning to work through the steps with prospects. You've started to establish yourself as their trusted advisor. You're qualifying them on their interest, money and decision processes, one intentional interaction at a time. And you have lots of different prospects with whom you're working. Where and how do you keep track of all this information?

Keeping track of multiple prospects is a work in process for selling companies and the vendors who support them. The most notable effort: software brands have created a number of tools, commonly called sales force automation or customer relationship management (SFA, CRM) software. Some of these are know by trade names such as Act!, GoldMine, Microsoft Dynamics CRM, Salesforce.com, Siebel and many others. They can be good, even great tools for containing information. They must not, however, be confused with a good selling process or great selling skills.

Alternately, you may work for a company that has a custom system, often referred to as a 'proprietary' system, to track contacts, opportunities and interactions with them.

On the other end of the spectrum, your company may have no tools at all. Although software tools have been available for over twenty years, an astonishing number of companies have never implemented or proactively managed technology to track contacts, customers, and interactions with them. There area number of reasons for this, problems which have not yet been successfully addressed by any software brand.

If you have software tools, we recommend that you talk with your technology staff to explore the feasibility of adding some data fields for the information we will describe below. Once the data is captured, it is usually possible to create a report or an information-rich dashboard.

If you do not have software tools, we recommend that your review the text below, and consider how you can capture the necessary information to support you well during the selling process.

There is no magic or perfect answer; but whatever you create will serve you far better that what you currently have. What is critical to Ultimate Selling is to know at a glance what you have and what you need from the prospect: What needs to be done to finish fully qualifying the prospect? What objectives from your selling process have been met? What still needs to be met?

This information—the who, what, when, why, how of each contact—is the information we need you to have available at a glance for each prospect. We call this the Prospect Profile.

This is valuable to you because you can open the file for any prospect, and know in an instant where you stand and what should happen next. From a foundation perspective, it keeps your prospects well organized. This alone will help them feel reassured by being well managed. From a selling perspective, it gives you a huge advantage: staying organized so that needed objectives do no slip through the cracks. From an accountability standpoint, your ability to state an objective and follow through on it tremendously underwrite being an advisor they can trust.

An Ultimate Selling Prospect Description holds a wealth of information at your fingertips during a sales call…and it serves as a kind of teleprompter, reminding you of what needs to be said or done during successive intentional interactions.

This is the chapter where you may start feeling overwhelmed. You've gone from the excitement of seeing how Ultimate Selling changes all the rules—in a commonsense, comfortable and practical way that works! You've been impressed enough by that to have worked through developing a SWOT about your company and what you sell. And you've even done a draft document of your sales process…and hopefully looked at our examples, and had another "ah-HA!" moment.

If you need a break before tackling this chapter, that's fine-take a break-but do comeback and acquire this.

Flex your muscles, and stay focused. You're becoming an Ultimate Seller. You've trusted us to guide you this far. If you've liked what you've learned, let's buckle up and finish strong.

The Prospect Review consist of eight sections or compartments for you to organize qualifying and activity information. They are:

1. Prospect Information
2. Current Situation
3. Desired Outcome
4. Consultant Information (if applicable)
5. Competitor Information (as applicable)
6. Sales Process Checklist
7. Qualification Status
8. Activity Logs
9. Post-game Analysis

Your assignment: scan over the table below. Keep in mind your requirements for selling your products or services. Don't get overwhelmed as this example can be revised *to just what information YOU need for YOUR type of sale.* While the sections don't change, the information you put in them could be just a little or a lot. Think of selling an office copier versus winning a contract to build a tunnel.

Table 14-1 The Art of Intentional Interaction Meets the Science of a Disciplined Approach

Section	Information	How It Works & What it Does
Prospect Information	Primary Contact(s) Secondary Contact(s) Title/Responsibilities Role(s): D = Decision Maker R = Recommender I = Influencer Stance: Friendly Neutral Unfriendly	Having to actually circle D (for Decision-maker) means you will have to *ask the question*: "Are you the person responsible for the final decision?" The form is your prompt to do it! Determining stance requires some tact: "I sense that you are uncomfortable with the amount of change this might require. Let's talk about that." Trusted advisor, one step closer.
Current Situation Source of their needs, pains and desired objectives	What is their current situation? What problems is it creating? What pain/s are they experiencing? Financial impact? Emotional impact? Other impact? Especially personal?	Jot down here the qualifying questions you know you especially need to ask this prospect, and what items you'll need to schedule for negotiation Will help you build an emotional ROI Their trust in your understanding helps you become their first choice for a provider.

Table 14-1 The Art of Intentional Interaction Meets the Science of a Disciplined Approach *(Continued)*

Section	Information	How It Works & What it Does
Desired Outcome	What are they considering as a solution? How will this fix it? What pain/s are they willing to endure in changing? What gains do they offer? Financial impact? Emotional impact? Other impact?	Gathering critical information here creates a bedrock for negotiation points—as well as giving you their specific wording for the Confirmation step
Consultant Information	Understand their role and agendas	Has this consultant recommended or used you before? If no, what's their role? Why are you involved—political, other? You will vet out their real role!
Competitor Information	Existing Vendor Other Competitors Information	Why would they leave them? Why are you there? With whom are you competing? What reasons? What is their perceived value?
Sales Process Checklist	Every action step that you can employ to make a sale happen… and make it happen!	What items from your typical sales process need to be included? What objectives? Which are unnecessary and can be skipped?
Qualification Status	Qualified on Interest Qualified on Money Qualified on Decision	What's left on your list of questions? What still needs to be negotiated?
Activity Logs	A journal of interactions and documented outcomes	What items and actions need to be documented? Keep records for yourself, and make sure you send documentation emails to your prospect!
Post-Game Analysis	Analysis of the outcome, won or lost!	What could have done better? What are you proud of yourself for accomplishing?

If you have SFA/CRM tools, many of the data points may already exist in your current system. Your focus should be on adding those that are missing and are needed for your selling process. What info does your type of sale need? And in your business, does the type information you need change from opportunity to opportunity…or does it tend to stay about the same?

Feeling overwhelmed? Take a deep breath. This isn't calculus. You'll survive!

Let's explore how some of the information or data points are the impetus for more questions. For example, I look at the location type (headquarters, region or branch) as a trigger for the more important question: do they have spending authority? If you are selling office equipment and they are a branch, they may need approval from the region or headquarters. I want to know if spending is centralized or decentralized. How? I simply ask.

Current Situation

This area supports qualifying on interest, where you document knowledge about what's causing them real problems versus their minor annoyances. Talking about stress, frustration, money, anger, loss of profit, headaches is critical. It's where your expertise at asking the right questions is critical. Do you have your master list of questions we discussed in the Chapter on Interest?

Listen to them carefully and compassionately. They are giving you real information about how to help them, and in so doing, you'll likely be in the lead to earn their business.

Documenting what you learn here will give you a powerful tool. Use what they've told you here when you get to in your Confirmation Step, to tie your solution of doing business tightly to their expressed needs.

Desired Outcome

Write down a list of what they need and want. This continues to support qualifying on interest as well as qualifying on decision.

If acquiring a new product or service they never had or used before, this information is the gap from where they are to where they want to be and what they are willing to do to achieve it. Just a quick list. Document, document, for future use.

Since you should have a pretty good idea of what your product and services cost, you can quickly determine if you fit the opportunity.

Include dates when product or service needs to be delivered and/or operational. Remember, once you've documented these, to confirm them via email with your prospect.

Consultant

There are some specific questions that are critical from an Ultimate Selling perspective, if you typically run into consultants:

Are you able to interface directly with the company (prospective customer)? If not…why? You must understand and a reason must be given!

Have they ever recommended your company, product or service before? If they haven't, this tells you a lot about the role you are more than likely playing on behalf of the consult. Are you a pawn in his game? If you're not sure, you must ask some additional, critical questions: "I'm confused. Since you have not recommended our company/product or service before..why are you involving us now?" Now the challenge of justification is on them, and you must determine if you continue in the process.

Competitive Information

This is another area where the rubber meets the road and your SWOT analysis can play a big role. Typically, the information collected should include a series of simple question in a few different ways. Some examples are:

- What other companies are you looking at?
- What other companies are or will be involved?
- How many companies besides us will participate in the process?
- I know the market and other vendors very well. If you would be comfortable sharing with me the names of the other vendors involved, I can save you some time and advise you if we should stay on your list and be involved. Fair enough?
- I can appreciate the fact that you want to keep that information confidential. However, I do believe in order to serve you best we need to begin with a certain amount of mutual trust. Do you agree? I'm truly knowledgeable about my competition, the marketplace and what our products and services can and can't do. If you can share, I may be able to save you a lot of time and effort by telling you where we might rank on your list…even if it means we should be taken off and out of participation.
- I would like you to feel comfortable in sharing our involvement with other vendors. Would you be comfortable sharing the others with me? If not, why is that?

This can sometimes be a sensitive and extremely touchy area for some salespeople. The reason we encourage you to include these types of data on your prospect description is because it allows you to learn about-and therefore gain control of—the competitor situation. You need to understand if there is typically someone that they're sharing this information with...it just isn't you! And you'd best find out why. Are you just the third proposal they require?

Do you really want to invest your time, money and company resources for a no-win outcome? This is all part of qualifying.

Qualification Status

Are they qualified on Interest? On Money? On Decision? Cross off what's done, and write notes below in your Sales Process about what needs to happen to keep them moving.

Your Selling Process

From your Selling Process, list or check off the activities that are critical to THIS opportunity. By having them marked, it's a visual reminder for you to follow your process.

Activity Log

Whether your prospect description is on paper, Excel or a SFA/CRM system, we strongly recommend that you *use it.* Intending to write something down but not doing so is worthless. Regardless of your technology or lack of it, it's critical to journal the events and activities you conduct with your prospects. Especially those that will make a difference in the sales outcome.

As we'll recommend in the next chapter, it's most important to never have a sales interaction without a stated purpose and objective, and to *always jot down what happened.* This may be very important later, especially in long selling cycles. It provides the paper trail you need to hold all parties accountable, including you. After all, someone must hold everyone accountable—and I'm willing to bet you feel you have the most on the line to win or lose. So, take personal accountability and do it. It's not something to delegate to anyone else!

Post-Game Analysis

Be straightforward and honest: it's essential. What was the final outcome? What can we learn to impact future outcomes? It may be just a quick note, but take the minute and reflect on it. We learn most, as you know, by our mistakes…if we let ourselves. Also, acknowledge what works well and keep doing it.

This is more of a management/data mining tool because an Ultimate Seller doesn't lose too many opportunities. That said, I suggest waiting about 30 to 60 days after the sale was concluded, in some cases after completion of the installation. Let the dust settle!

The objective is straightforward and simple: learn everything you can about what went right and what went wrong, and take that information into future thinking and action to improve outcomes. It's no different from sports when they study game films to find out how they can be better in the next game.

If you are unflinchingly honest with yourself, you will learn about your skills in the selling process, and you'll find yourself more naturally improving your abilities and sales results.

Summary and Assignment

We understand that this may be more documenting than you have ever done in your life! If you do large sales with a long cycle, we cannot urge you strongly enough to establish this best practice. If you do small sales with a short selling window, it may be something that you don't believe or feel should be committed to paper. My position is, if you're losing sales, try it. What do you have to lose—except more losses? And much to gain!

Take a few moments now and sketch out a Prospect Description. or enhance your existing one as an Ultimate Seller. Even if you have a software system already designed, it may not reflect some of the concepts presented here. Fortunately, most are easily modified if you can present a compelling case to your staff.

As with all prior exercises, your effort will be pay off handsomely as you will be able to use this same structure on all sales interactions.

Regardless of the medium you use to keep this information—database, report, Word, Excel, paper, whatever—you will need to refer to it when you are interacting with the prospect, to keep you focused, on track and ensuring you capture all the important information you need to continue the sales process with them.

Chapter 15

PLANNING YOUR INTERACTIONS—INTENTIONALLY!

Planning for each interaction sounds laborious. It sounds like even more preparation, doesn't it? Gasp! *More*???!?

Yes, it is; but relax. It's easy.

In the last three chapters we've been working a lot on preparation from the KNOWLEDGEMENT MANAGEMENT perspective:

- Knowing what your selling process is
- Knowing what your objectives are
- Knowing your SWOT

Now let's return to working on preparation from an ACTION perspective. You've prepared. It's time to circle back to where we left off earlier—qualifying, negotiating, confirming and connecting.

As we talk about preparing for intentional interactions, it's important to remember that even though people do consider price and product/service features when making buying decisions, it is still powerfully true that people still buy from people they like. Even more, they buy from those they trust. Despite the speed and ease of texting, email and evolving technology options, it is hard to build a relationship of real trust via text and email.

Personal interaction counts. The more important the interaction, the more important handshakes, eye contact and the human voice are. So as you plan interactions, give thought to which are of minor importance and can be email, text, tweet, etc, and which should include a more in-person approach.

Next, some clarification. When we say "call" in the following text, we mean any interaction, whether in-person meeting, internet meeting, telephone call, conference call, letter, email, text, tweet or Facebook post. Regardless of the medium you've decided is appropriate, it's happening because there are clear objectives that must be achieved.

The word 'objective' may sound familiar. Remember when we asked you to create your selling process? Remember that there was a column entitled "Objectives"? Some objectives, like those in your selling process happen in every sale. Some are unique to each sale. Regardless of kind, they form the basis for planning your "interactions agenda". These are the hurdles you need to clear. They also form the basis for the questions you'll need to ask in various calls.

We're going to go over some specific guidelines about interactions (calls). We will review

- The game rules
- The strategies—the do's and don'ts of interpersonal selling activities

Game Rules

Rule One: Everything for a reason

We never engage with the prospect unless we have a clear reason and specific objectives. If you could define the best possible result of your interaction—you got the appointment, you qualified their financing eligibility, you learned their biggest worries—that's an objective. Every time you interact with your prospect, there should be a reason for it, an objective you want to accomplish, and a plan for how you're going to make it happen. Make it a priority for yourself to be consistent about this.

Rule Two: Purpose trumps style

It's okay to appear informal. It's okay to be conversational. Or you can be formal. Whichever suits your personality, remember that even when you may appear to be entertaining, your purpose is to advance a sale.

Rule Three: Face the mirror

When the interaction is over, assess honestly whether or not you achieved your objectives. If not, what did you miss? What went awry? When did you lose control? How? What must you achieve on the next interaction?

Rule Four: Keep it clear

Once the interaction is over, summarize and confirm it with the prospective customer via email what was accomplished, what was agreed to, what should happen next, and by which party. Identify any open, unanswered items that came up.

Follow-up documentation via email of each interaction prepares a foundation for your next objective. These emails send a great message in a subtle, yet very clear manner:

- It defines the future action plan—which keeps you more in control
- It has the reassuring effect of having your prospect feel each event is well managed
- It demonstrates personal accountability…on both sides. You most definitely want that from your prospect before you continue to invest time in working towards a solution with them.

Here's an important question: if your prospect fails to demonstrate personal accountability, what insight will that provide you regarding them and their "opportunity"?

Rule Five: Go Step By Step

Regardless of whether you have a complex objective which requires multiple interactions, or a simple one, a clear outcome in which everyone knows where they stand and what the next steps are, is critical before qualification can continue with the next objective.

In your selling process, you also listed mechanisms: tools and action items which are resources to support achieving those objectives. You listed people who are selling resources, and people who are important to your objective but may or may not be supportive.

Planning for each interaction must be deliberate and intentional. Using your selling process, you are going to build a specific plan for each objective. Every interaction with them will serve a purpose. This is your agenda. Take your specific objectives, plus people, plus mechanisms and create specific, detailed plans to accomplish objectives quickly and clearly. This becomes your agenda.

Rule Six: Keep Your Enthusiasm In Check

Please note that when you are in your sales interactions, you are not, not, *not* to hear their remarks and get overly excited and immediately jump into presentation mode and quickly let them know how your product/service will fill their needs. NO "have I got a solution for YOU!" temptations.

Let them figure that out with you as you go along. You're simply going to ask more and more questions, taking notes.

You're going to stay in control. Every interaction should be geared to helping you validate and qualify the opportunity further.

As a way of visualizing these plans, we refer you to Figure 15-1, 15-2 and 15-3. Feel free to modify them for your particular selling process.

Only then will you then be prepared to calmly assess where your prospect is coming from and apply a Go/No-Go determination (as illustrated in previous chapters on qualifying for Interest, Money and Decision with a green, yellow and red-light metaphor.) These lights are a 'heads-up' that you need to ask additional questions—and drill down to uncover issues that will be negotiated and converted to green.

Before we begin with the objectives, let me take a few moments to remind you that good effort in each of these activities will pay off in spades. Your first efforts here are the biggest part of the work. Once you've done them future sales activities will need only limited modifications. In essence, you will have a very high ROI for your time and effort to complete the assignments here.

Figure 15-1 Interaction Planning—Overview

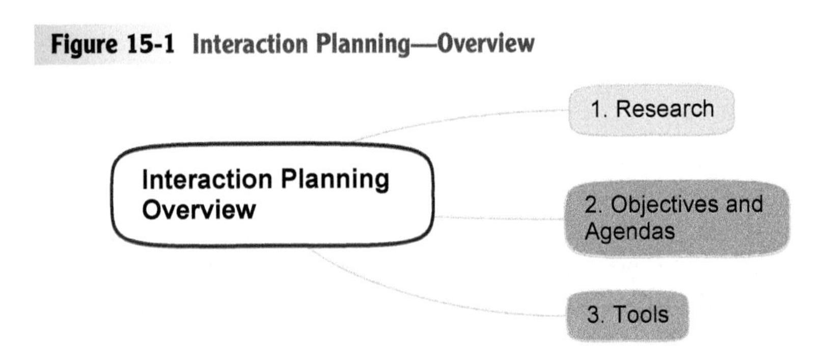

Figure 15-2 Interaction Planning—Research & Tools

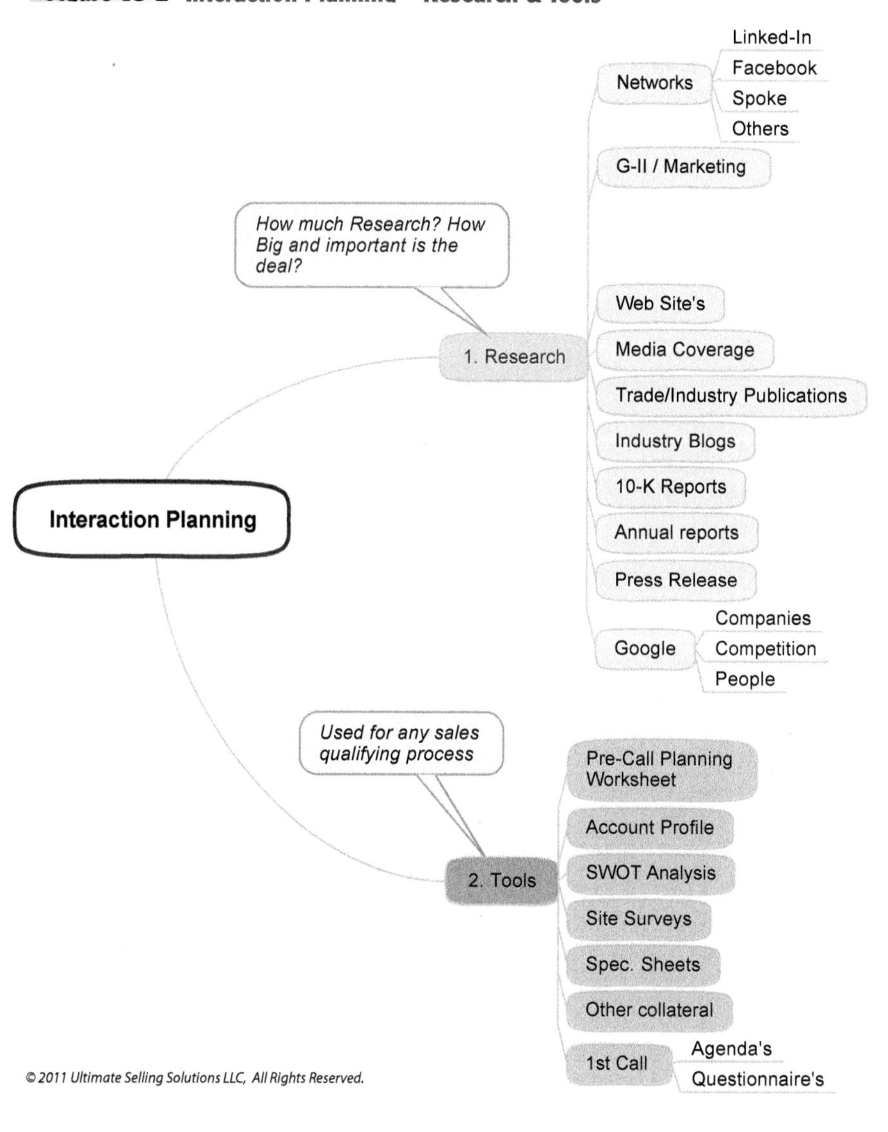

Figure 15-3 Interaction Planning—Objectives & Agendas

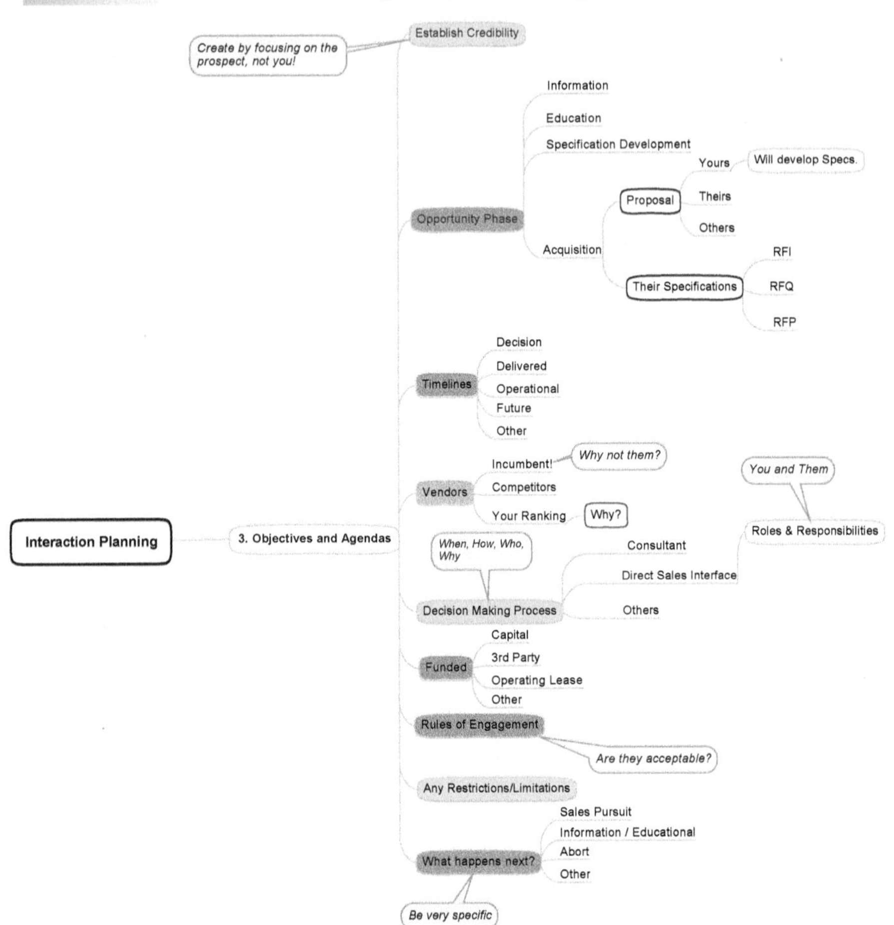

Interaction Strategies

The simple, strong strategy of your process is to take control as early as possible, in a respectful and intelligent way.

Strategy: Ask specific questions about the project

Strategy: Be honest if you are nervous. Nothing clears up nerves better than simply and honestly saying so. "I'm always a little nervous getting started with questions. I want to do a good job, and I worry that I'll forget something. So please let me know if you think we need to cover something more deeply."

Strategy: Don't ask questions that can be answered with a yes or no. Ask open-ended questions that start with who, what, when, why, where, how.

Table 15-1 Control the Conversation by Asking Better Questions

Not so good	Better
Do you have a budget?	How have you decided on your budget?
Are you able to increase your budget if needed?	If in talking we find that your budget needs to be increased, what's the process for that?
Questions that start with "do you" "is there" etc.	Questions that start with or include "how" "what" "who"

Strategy: Since we are just beginning to qualify them on Interest, Money or Decision, we don't know if they are a good opportunity. No matter what the prospect thinks is the priority, our priority is to Qualify on Interest, Money and Decision. Keep the conversation coming back to these three subjects.

Strategy: Prepare yourself to handle contingencies. What if you learn in the conversation that they're really not interested? How will you end the conversation gracefully and leave future possibilities open? What if you learn in the conversation that the opportunity is bigger than anything you've taken on before? How will you respond?

What if you learn that their interest is uncertain? What specific questions will you prepare to find out why? Have they been burned by prior projects? Are they afraid it will cost too much?

Strategy: To draw out more information, Use "active listening" to draw out more information. Paraphrase what a prospect says and repeat it back. The effect will cue your prospect to expand with more detail on what they said before.

It also have the effect of them realizing that you really are listening and that you care about what they are saying, without trying to rush them to a sale. It's a rare quality, and very attractive to prospects.

Summary

You won't get it all right away. But with practice you *will* get it. The critical key here is that you do not go into any meetings with prospective customers without clear objectives and a clear agenda. While it may appear to be very relaxed—and that's not a bad thing—it's not a social situation and it's not a car without a steering wheel.

Your interactions with prospects when you operate from a planned, prepared perspective says that you're professional, that you follow up, that you stay on top of things and are trustworthy, and that you're interested in protecting them and their interests. It's a very big piece of the puzzle in becoming a trusted advisor.

Chapter 16

Real-Time Review

We said earlier that you either have a plan or you will be part of someone else's plan. In Ultimate Selling, it all started with your preparation work. You established objectives and crafted an agenda with planned interactions to qualify your prospective customer on Interest, Money and Decision... as viewed from their side of the desk, not yours. Remember: it's all about them!

You did not just show up at the account to see what happened. You had a plan. So did your prospective customer. We need our prospect's plan to either match our plan, or carefully work with them until it does. Remember, maximum control of variables is a tenet of Ultimate Selling, accomplished by your direction and re-direction until the prospect discovers it's their plan.

We've gone over the secret to adapting their plan to fit your plan, suggestions and ideas built on honest and compelling reasons that would best serve *their* interests, not yours.

How do we do real-time, in-progress assessments on how we're doing with a prospect? What's first, second, next?

- First is an absolutely unflinching, honest *assessment* of each interaction, per rule #3 of Intentional Interaction.
- Second is being honest with ourselves again: do we believe this opportunity still *merits our pursuit* of it? Or not?
- Third: if it merits pursuit, how should we refine our plan, if at all? If the outcome was negative, can we get it positive again? How? If the outcome was positive, how can we build on that to increase our position? Who do we need to talk with? What's our avenue of interaction? What's our objective? What can we negotiate on further? Remember, it always has to be a win-win.
- Fourth is to implement the revised plan.

Interaction and Skills Assessment

How well did the interaction go? Did you accomplish your objectives? If not, what went wrong? What would you have done differently?

Pursuit Decisions

Honesty with yourself, especially when deciding whether the prospect is worth going after, is critical. Under what circumstances does it make sense for you to pursue the opportunity? It might make better sense to decide that you are *not* willing to pursue it any more. If you need, get the advice of others to help with that decision.

If you decide that you aren't going after the opportunity, you take it away from the prospect: you inform the prospective customer of your decision to walk away. If they fight hard to keep you in the process, they will need to meet your conditions, or you stay your course.

If you decide to continue working with them, you'll let them know what conditions need to be met, from your perspective.

We're talking about conditions you are confident will result in a high probability of success...conditions that must be met to keep it a win/win.

We're talking about facts, actions and events...not hopes and wishes, or the fact that you may not have any other prospects to pursue. We are Ultimate Qualifiers, and are not afraid of walking away from an opportunity which won't succeed.

Anything short of mutual respect that serves both buyers and sellers is a win/lose scenario. A win/lose scenario is never one an Ultimate Seller would pursue. It cannot result in Connecting, and in the end, inevitably becomes a Lose/Lose.

How to communicate your decision not to continue

When you decide that continuing requires certain conditions to be agreed upon, *you* will (as usual) use interaction planning to prepare for it...and this will not be an email conversation or negotiation!

You will present your conditions as serving their best interest. If unable to obtain agreement from them, you will state that as an expert in your field, you could not in good conscience participate in a process that, based on your professional experience, would negatively affect them as a potential customer, and indicate that you respectfully intend to withdraw.

Talk about expressing confidence and creating insecurity in their minds! I can promise you it rarely happens that a prospect hears a salesperson stand up and withdraw. To some of you it may sound harsh, but I promise you it will make them think, and can most definitely change the dynamics in your favor. If it doesn't, they do not meet the criteria of being qualified on Interest, Money or Decision. or Money, depending on the issue—and I'm also confident you were never really in the game anyway. Now you won't invest your time and resources in a lost cause—but you decided that on your terms and early in the game, not on theirs.

As stated earlier in this chapter, if they fight hard to keep you in the process, the pressure and responsibility will be on them. You just need to hold your position and know exactly what terms and conditions you require to continue, plus their commitment to act on each of them, giving you now a high probability for success.

Revising Your Sales Pursuit Plan

Wanting success is not enough. You need to adapt your plan as required. You've built your intentional interaction plan from your selling process objectives and your prospect description. Now you need to revise your plan to build a bridge from where you are today to where you want to be.

Having a defined yet flexible plan allows you to continually assess where you are, decide what you need to do next, and make the necessary revisions to stay on track.

Simply said, plans do work! Having your roadmap for success will keep you from getting lost. Your requirement as an Ultimate Seller is to have the confidence, courage and willingness to press on to your ultimate destination: a successful sales outcome.

While not all-inclusive, we have provided a Mind Map example to assist you in assessing call outcomes and revising your intentional interaction plan. Refer to **Figure 16-1—Post Call Analysis & Sales Pursuit Plan** for observation and further exploration:

While we have described Ultimate Selling in many ways, it's very much based on keeping an eye on that critical crossroads where you either fall under the spell of your prospect, or whether you move forward in charge,

Figure 16-1 Post-Call Analysis & Sales Pursuit Plan

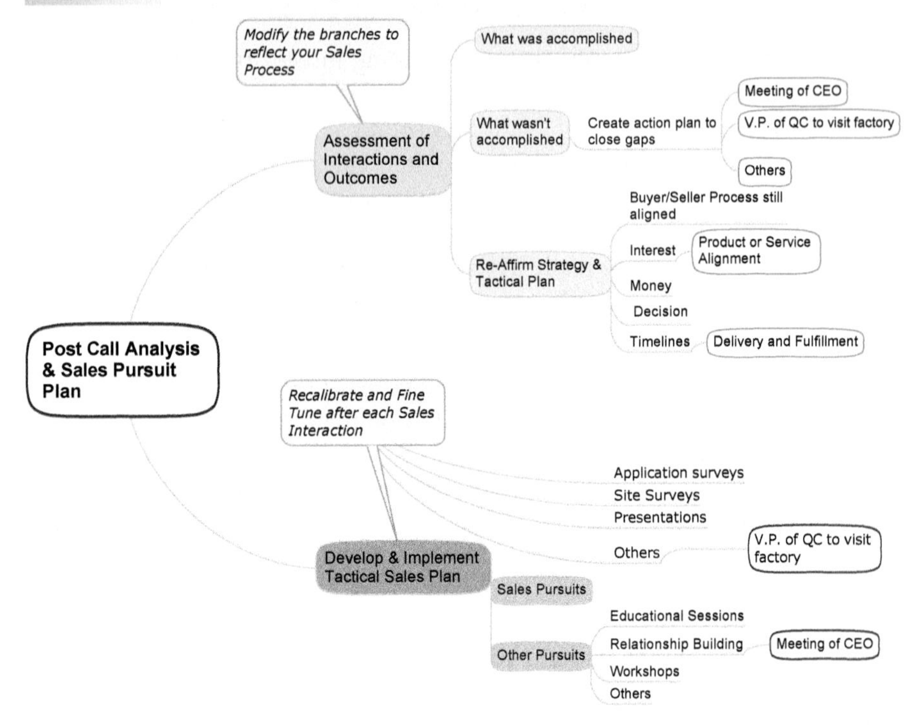

with maximum control, planning the course which you believe to be a successful route.

So no falling under spells! Review your plan, decide where you are today and what's been accomplished so far, assess the gap and adjust your plan with a strategy to close that gap.

Remember, you are only going to plan activities that address the interests and concerns of your prospective customer, and the ones you may require for a successful outcome, nothing more and nothing less. Don't go crazy; only implement activities you genuinely believe will achieve your objectives.

Before you proceed to the next chapter, take some time to mull over what we have just covered.

Chapter **17**

Sales Forecasting or "The Emperor's New Clothes"

Traditional sales forecasting tends to be more speculation and hope than a solid expectation based on real actions. In fact, organizations have become so lax with accurate forecasting of sales that they have created the "rolling forecast".

All too often that hopeful forecast doesn't actually close, but the hopes continue. The forecast report has really become a pipeline list of unqualified prospects, or worse, leads and suspects. The common, albeit incorrect 'fix' is to simply roll the forecasted sale forward to next month rather than recognizing symptoms of the bigger problem: many organizations have woefully inaccurate forecasts, because their salespeople don't know how to qualify and forecast accurately.

Do you really want to put your financial security in a rolling forecast based on speculation and hope? If you do, it's a disservice to you, your family, your peers and the company.

In too many cases, opportunities that are forecast as sales have little to no chance of converting to a sale...but it's seen as a better alternative than having to forecast nothing! Can you imagine how your sales manager and organization would react to your saying you have no prospects expected to close in the next 30–60–90 day window? They are afraid of seeing the truth! The real truth is that they'll have to deal with it one day anyway, when those hoped-for forecasts fail again to materialize.

Here again is a symptom of the same big sales problem: hope and fear are driving business actions (and worse, business planning), instead of intelligent strategies designed to result in profitable outcomes.

Organizationally, this can be disastrous because it provides an illusion of forthcoming business. This can impact product availability, service and support, manufacturing and material lead times. and/or staffing requirements. Personnel resources may be inadequate, excessive or misappropriated. In the words of Wall Street, it's "missing your projected revenues and earnings".

Let's look at an example of some of the impacts to an organization when sales forecasts are wrong and hoped-for revenue doesn't materialize:

Table 17-1 Effects of an Inaccurate Forecast

Who is affected	How they're affected
Salespeople	Can't pay your bills with unearned commission or bonuses, and personal debt may increase if you're spending future "income" on today's lifestyle Employment may be in jeopardy Domino effects to your well-being, and that of your family and friends.
Sales Management	Ditto the salesperson above, plus: The stress of being the middle 'vice" between their sales team and upper management Evidence that they can't manage
Upper Management	May decide to do major capital spending to "prepare" for increased anticipated sales May hire personnel to ramp up, for same reason May be personally accountable to stockholders when company fails to perform and expenses exceed revenues
Organizational	Loss of trained and skilled personnel that may not return for various reasons when the business improves/recovers Bad attitudes organization-wide Stock value declines

An inaccurate forecast can be wrongly pervasive across a company!

Forecasts of new business prompt companies to ramp up in staff and equipment in preparation for an increased workload. In many cases, that's not an overnight thing, and is done via expectations set by sales forecasts. But if the new business isn't realized, the new folks hired will quickly lose their jobs. How would you like to be in those shoes…or the person telling them?

When big investments such as new trucks aren't really needed, the entire company can suffer financially. It may be your base pay that gets cut, or your sales position, if they close your department due to significant losses. Then it's your security and financial future on the line.

It sounds drastic, but in truth, that's the kind of planning and buying decisions that can result from poor forecasting.

Many sales departments and their upper management have tried to address this problem by weighting forecasts. In either case, the individuals involved and the organizations they work for are not addressing the real issue: the need for more accurate forecasting.

Want an opposite scenario? Under-forecasting is just as bad! Increasing sales without the company having a heads-up can result in supply and human resources shortages. This leaves your customers with an unhappy experience. What you promised them doesn't get delivered or supported as they expected. Inaccurately low sales forecasts are just as bad!

Back to you and you alone: in the world of professional selling the sales forecast is all you have as a measurement for whether you are succeeding or failing at selling. It's an important gauge for your future financial security. As an Ultimate Selling professional, you want to know exactly where you stand with your prospects. How else can you quickly identify any discrepancies between what they want and what you want, be proactive and take corrective action?

Granted, sometimes various nuances may apply but, I believe you will find the methodology defined in Ultimate Selling will get your forecasts cleaned up and accurate in short order. And you will know with a high degree of accuracy when and where your sales objectives will be achieved.

In Ultimate Selling, it's all about reality! You must have the honest ability to assess where you are early and often so that you can be proactive in creating solutions to overcome challenges in a timely manner. The last thing you need is a rolling forecast with a perception that things are great. This is where you have to "man up" or "woman up". Scary and intimidating? You bet! Again, imagine honestly projecting NO SALES!

Think about your last six to twelve months. How many of the sales you forecasted actually happened, even after you rolled them forward several times? How much money did you actually earn, compared to what you'd hoped to earn?

How would it feel to get that turned around? To see more and more sales in which you've invested significant time closing? To have meaningful interactions with prospects, instead of leaving voicemail after voicemail, unanswered email after another…or to put an end to forecasting opportunities at 85% because you were requested to make a proposal—but for which you rarely got a check because of problems at the close?

Get this radical concept through your head:

- a sales forecast is NOT a weighted pipeline. Repeat: A SALES FORECAST IS NOT A WEIGHTED PIPELINE. We repeated that, louder, because we believe that in many companies, the pipeline, weighted, has replaced a true forecast. It shouldn't!

- a forecast is NOT simply a list of prospects who are in the process of being qualified
- a forecast is not the same as a funnel

So what is a forecast?

A Ultimate Selling sales forecast is you committing that this sale IS going to be resolved (won or lost) in the 30/60/90 day period. The prospect and you have discussed it and you are in agreement about this.

While pipelines and funnels have their place, they are not the same as forecasts. They are, instead, the supply lines to forecasts. They are your leads, they are the folks you are talking to the first time, they are prospects in the initial stages or later stages of being qualified with whom you're working on negotiating win-wins on options.

Your sales process and subsequent intentional interactions are your path to weeding out unqualified prospects from your forecast. Only qualified prospects earn the right for using your time, money and resources in pursuit of a sale with them, and the right to be included on your forecast. The keyword is sales "pursuit"—you're not giving your prospect a free education—you're pursuing a sale.

There is a distinct line between your pipeline and your forecast. An opportunity from your pipeline gets put on your forecast when it reaches the 70, 80 or 90% likelihood of closing within the next 30, 60 or 90 days.

Sales forecasting as defined in Ultimate Selling is not speculation or hope for a final outcome. Rather, it is a calculated outcome based on information provided by the prospective customer—their right, as you mutually see it—to reside on your sales forecast in the likelihood that your relationship will connect shortly. In Ultimate Selling, there are several expectations.

EXPECTATION 1: A sale is defined as a completed opportunity that has all the required documentation—purchase orders, executed agreements, required deposits, financial approvals, permits, authorizations, etc.—in order for your company to process the business. It's not a sale until it's done!

EXPECTATION 2: Sales in the 30-day timeframe will be concluded, whether they are won or lost, within the designated time…otherwise, they don't belong in the 30 day timeframe. Once they are listed in your 30-day forecast, they must be removed from the next sales forecast. Period. No rolling allowed.

Note: For rare cases with exceptional circumstances, they may carry into the Aged section for one month if not resolved in the designated time frame of the current 30 days. For example, there was a flood at their main office. They can be moved to the Aged section for no more than 30 days. It's a reprieve for a limited period of time.

If they still don't close, they are then removed from the next forecast. Although at one time they may have been qualified, the circumstances 'unqualified' them...so you now go back to qualifying them if and when circumstances dictate.

EXPECTATION 3: You should be working with enough prospects to always meet or exceed your revenue objective. (In our sample, the quota is $200,000 in monthly sales.)

EXPECTATION 4: You will maintain enough prospective opportunities in your 60 and 90 day timeframes that will progress and sustain your 30 day quota requirements. (Remember, again, that your 60 and 90 day time-frames are NOT a list of prospects that you plan to push forward into a sale in 60 or 90 days. They are a list of prospects that you have a high expectation WILL COME TO FRUITION in 60 or 90 days).

We believe that nearly all sales transactions in the final buying phase fall within a 90-day window or less, once a critical amount of information starts being exchanged between prospect and vendor(s). Therefore, an Ultimate Seller's forecast does not project beyond 90 days, as anything beyond that is too speculative and they simply belong in your pipeline.

If the products or services you sell are always shorter and faster sales, you will want to adjust the time frame accurately. Just make sure that your forecasts are accurate within those windows.

EXPECTATION 5: Regardless of your company's current method for sales forecasting, you will integrate the principles of Ultimate Selling into your existing system (or create a new system which incorporates our principles for your personal use). As we will discuss later in this chapter, this tool will absolutely support positive sales outcomes—and more confirmed sales are worth the extra work.

EXPECTATION 5: Accuracy is expected, meaning you will know the specifics of When, Where, How and by Whom each sale will be concluded

EXPECTATION 6: Honesty—with yourself and your sales manager—is expected. Honesty is crucial.

Let's continue with sales forecasting the Ultimate Selling way by describing requirements for an opportunity to be entered into the sales forecast.

EXPECTATION 7: The prospective customer has been qualified on Interest, Money and Decision. They've committed that their buying timeline is within your 90, 60 or 30 day windows.

EXPECTATION 8: In Ultimate Selling, anything outside of the 90-day window means they *are not* qualified on Interest and/or Decision and/or Money yet. Keep them in your database, and keep working to qualify them...but keep them off your forecast!

EXPECTATION 9: Accounts and opportunities will enter your forecast at their own true level (for example, they may enter as a 30-day rather than a 60–90 day!) and will progress at their own speed—but they must always qualify on Interest, Money and Decision before they are allowed on the forecast.

EXPECTATION 10: Move, or get the boot: *under no circumstances* can they remain passive in the sales forecast or the sales process. They must keep moving through the process until won, lost or aged, after which they are off the forecast! Stagnant 'opportunities' are like standing water: it does not take long for it to go bad. You have to keep them moving to keep them fresh.

EXPECTATION 11: You will know in the beginning of the month or quarter if you will have sales success, rather than after the month or quarter. If you find you're having different results at the end of the month or quarter than you expected 30 days out, quickly identify and proactively address where your process isn't working well.

EXPECTATION 12: It's possible for an opportunity to be removed from the sales forecast based on no action, and then at some date reappear on a future forecast. (for example, they were ready to issue a purchase order but due to a unique circumstance the acquisition was put on hold indefinitely. Approximately 4 months later you were notified that they are now ready to proceed again. Now, you need to re-check qualifying and determine whether they are in the 90, 60 or 30 day window.

While this is a great tool for the sales manager, it's even more important to you as a sales professional—because it's your personal measurement tool on predicting success and winning business. As you continue to work on Ultimate Selling skills, it will give you some metrics to see your improvement over time.

Figure 17-2 90-Day Sales Forecast

Report Period: 8/1/2010 through 8/31/2010
(Insert personal information for Territory, District, Region, Division or any other required company identifiers

	Account Name	Description	Dollar Amount	Rating 0-10	Notes
1	ABC Company	Model AZ213 Heat Pumps	$57,298	9	PO in process, due 8/13 Begin installation 8/25
2	US Tool Company	40 kw Turbo Diesel Engine	$26,789	9	Have executed agreements,
3	Diamond Builders	HVAC Units	$365,000	10	Have PO, Deposit $ and need permits, due 8/15
4	Shaffer Contracting	HVAC Units	$126,000	8	Gave verbal but my gut has concerns. Insist it ours
5					
30 Day TOTAL			**$575,087**		
1	USA Contractors	15,000 Watt Generator	$13,248	7	Awaiting Leasing approval
2	Johnson Builders	HVAC Units	$82,000	8	Deliver by Sept. 25
3	ZZZ Builders	HVAC Units	$142,788	9	Received verbal for PO & Deposit check by 9/15
4	Main Street Builders	Model KX111 Heat Pumps	$68,298	7	Awaiting bank funding for Phase II
5					
6					
7					
60 Day TOTAL			**$306,334**		
1	Falcon Builders	HVAC Units	$48,000	7	Needs by late October. 2 other vendors being considered Trane & Carrier
2					
3					
4					
5					
6					
7					
8					
9					
90 Day TOTAL			**$48,000**		
1	Kenilworth Builders	HVAC Units	$267,936	9	Need Executed Agreement. Requires president's signature and there was a death in his family. Should acquire by 8/10
2					
Aged TOTAL			**$267,936**		

Rating Your Likelihood of Success

Let's discuss the most critical data point of an Ultimate Selling Forecast rating.

In Ultimate Selling we consider the rating column to be critical in deciding their probability. On a scale of 1 to 10, 10 being a closed sale, it shows where we stand with them and where they stand with us. The notes say what needs to happen next to get to 10. Let's look at our exhibit and review how we use ratings.

Scenario 1: Rating a forecasted sale as 10 means that they have a 100% right to be on your sales forecast, expecting to close within 30 days…and you are 100% confident about delivering their business to your company, and your solution to their company. You know that the opportunity is solid because you've *fully* qualified them, and there is nothing left to be negotiated except the action steps to obtain the executed agreements, collect funds etc. You plan to be doing a Confirmation, or have completed a Confirmation step to be followed by Connecting, *this month.*

You've *also* asked the prospect very clearly and honestly about their perspective on this: "Michelle and Gian, it feels to me as if we are very close to an agreement on working together. I'd like to know where I stand from your perspective. On a scale of 1 to 10, how likely are you to sign/execute an agreement this month?

It takes guts, but it's effective. In fact, most of the qualifying questions in Ultimate Selling take guts because they are so honest—but they do work!

Scenario 2: You put the opportunity on your 30-day timeline rated as an 8. While you expect to be Confirming and Connecting within 30 days, work still needs to be done.

Why aren't they a 10? They may be fully qualified with a few minor points to negotiate, or it may be that the problem is on your end. You may not have the ability to get shipment as fast as they want. In this case, it's your responsibility to close the gap between 8 and 10 by clearly describing to yourself and your prospect what the issue is, and negotiating a win-win with your prospective customer. You expect to conclude those negotiations within this month, as well as Confirm, Connect, and have all documents/payments processed.

What if you thought everything was fine, but when you asked the prospect about their perspective, Michelle replies to your question this way: "Well, Jim, I'd say we're about a 7 on the 1 to 10 scale. Maybe 6."

How do you close that gap? It's simple, and it's the way you've worked to this point. You explain to them that you'd like to get to a 10 in their eyes, and ask what you need to do to move to that number.

At this point you will determine what stands between you and the sale. And, typically, you will learn whether or not you have the ability to move it to 100%. Yikes! that sounds like we weren't talking to decision makers! The good news 'is that' you can now address it proactively rather than trying to reverse a decision that did not go your way.

You can take it a step further with the prospective customer, to reaffirm and validate what was stated. It's a warm-up to Confirming. In dialog, it would sound something like this:

Michelle, let me see if I clearly understand what you are saying. If we could satisfactorily address item A, B and C, does that mean we will definitely earn your business?

If they answer yes, you follow through with negotiating A, B and C. If they answer no, you find out what else the problem is. They may not make your forecast, but you're back to qualifying and negotiating to drill down, and you have more insight as to what's affecting the outcome. If they won't answer questions, they are hedging on something…and your challenge is to uncover the reason(s).

In Ultimate Selling we consider anything less than a 7 to be an unqualified prospect. You must never allow anything less than a 7 to be on your sales forecast, for any reason. Keep them in your pipeline. keep qualifying, keep negotiating and find out the hidden agendas. Are they using you to get an education? Price checking against another vendor? Forecasting them is premature!)

What about the Notes section?

The notes are a 'last chapter' of your selling process with this prospect. Besides objectives, write down your comments and thoughts about anything you had determined could be helpful at some point in securing the sale. Either the prospect told you or what you already know. It may be moving an 8 to a 9 or 10. Keep these notes handy so you can use at the appropriate time as the sales process progresses.

Ratings Provide Great Insight on What Happens Next!

One of the critical elements of ratings is to be honest. You don't have to like the rating; you just have to be fairly accurate. Ratings are good things. It's where you become serious about your outcomes.

The Ultimate Selling sales forecast is a window into reality. It will tell you about revenues, personal income and degree of success. Conversely, it will also quickly identify what's not happening and what needs to be done to reverse the circumstance. It's your reality check, like it or not.

Let's take a quick look at the sample forecast provided and see what you can learn from the 30-day window of your sales forecast. If you slip into having it become an extended pipeline or a rolling forecast, be firm and remind yourself: it's an *Ultimate Sellers* 90-day sales forecast.

Do you have enough opportunities concluding in the 30-day window to achieve your monthly quota?

- If I only get Diamond, I will significantly over-achieve!
- Diamond could slip into the Aged section if I don't get the permits. They're a critical activity for which I must schedule time!
- Shaffer could surprise me the wrong way, and be a no-deal
- Lose Shaffer, Ritter becomes aged, and I will not make my quota. So I need to make sure US Tool and ABC company close as well. While small, they make a difference.
- I should be able to confirm Kenilworth from the Aged section carried forward from last month since the president has returned from the family funeral. If all goes as planned, it should be a banner month!

Do you have any questions about what action steps to focus on, to ensure these deals? In our example it would be 'get Diamond's permits, keep close in conversation with Shaffer, and re-contact Kenilworth's buyer. Stay on US Tool and ABC Company to get the PO completed.'

What does your OWN Ultimate Seller 30-day forecast tell you that you need to do? And not at the end of the month, but now, day one? Strategizing and defining tactics to improve your sales forecast outcomes should be your immediate and most important item on day one of each new month.

What about the 60-day window? What does it tell us?

- Look at how the 60-day forecast stacks up for a good 30-day the following month.
- In the 60-day or less sales cycle, even if you get your most probable, ZZZ Builders, if nothing else closes you will be well short of your monthly quota.

- I have a critical need to build my pipeline for upcoming months and quarters!
- What do the notes tell you needs to be done? Don't you wish the person who wrote the forecast (what if your boss handed it to you?) had given you some specific points to address? Now you're thinking! Do the same in your notes, so you can stay on top of everything!

And the 90-day window?

- You'd best start aggressively prospecting or mining your database, as it's clear you do not have enough serious opportunities in the 90-day window to succeed in the forthcoming 60 and 30-day forecast!

In short, you will learn all you need to know about what must happen next in order to succeed. And you will most definitely have insight into what's going to happen as it stands today, good or bad, which you can use to take action on the required activities.

This practice of being honest with yourself about what you've got—and the opportunity it gives you to improve things—reminds me of a poem I've had for many years and would like to share. It's titled ***"The Guy in the Glass" by Dale Wimbrow:***

When you get what you want in your struggle for self
And the world makes you king for a day
Then go to the mirror and look at yourself
And see what that guy has to say

For it isn't your father, or mother, or wife
Whose judgment upon you must pass
The fellow whose verdict counts most in your life
Is the guy staring back from the glass

He's the fellow to please—never mind all the rest
For he's with you clear up to to the end

And you've passed your most dangerous, difficult task
If the guy in the glass is your friend

You may be like Jack Horner and 'chisel' a plum
And think you're a wonderful guy
But the man in the glass says you're only a bum
If you can't look him straight in the eye

You can fool the whole world down the pathway of years
And get pats on the back as you pass
But your final reward will be heartaches and tears
If you've cheated the guy in the glass

If you live by the creed and tenets of Ultimate Selling, you will not cheat yourself, but will instead have a life of serving yourself and others well.

Summary of Critical Points to Ultimate Selling Sales Forecast

- The input is acquired from prospective customers, not your opinions
- Honesty is paramount or you will be cheating yourself
- It will identify what specific actions are required for successful outcomes

Chapter 18

GAME ON: LET THE SELLING BEGIN!

Congratulations! You made an investment in yourself with sweat, time and a little money for the book and ground your way through these chapters, and you're about to embark on a new journey. The sales methodologies you used before will increasingly become part of your past. ***It's time to take the field with Ultimate Selling.***

As you know from this chapter's title, selling is a game. Like sports, you win or lose. Draws are not applicable: if you're willing to settle for a draw, you were never in the game.

How many times have you missed being in the real game—a great opportunity—because you are so busy investing time and resources chasing another opportunity and never even got on the field?

Of course, you can't always know about all the deals happening, but you can improve your odds if you're looking for the games you want to play, rather than hoping to get into all of them and ending up sitting on the sidelines with all the other players doing the same.

In professional selling, we refer to it as the ***Controllable versus the Uncontrollable.***

I have a very high degree of confidence in Ultimate Selling. With our approach and rules of engagement, I would expect to win all the games, were there no uncontrollable factors. We'll never get it down the uncontrollable to Zero, but that's the type of circumstance we want to work towards by using Ultimate Selling.

Regardless of whether you sell items for millions, hundreds of thousands, or only cost a few hundred dollars with short or long sales cycles, you can do all the right things and an uncontrollable can come in to end the game. You're scheduled to pickup a check and the buyer is out for an extended period with a serious illness. Your chief decision-maker decides to retire or take a leave of absence, the company's finances take a terrible hit from a lawsuit...it happens. Vince Lombardi said, "we didn't lose the game; we just ran out of time". Vince's comment is too true in these cases!

But barring those uncontrollables, bringing the rest of the game to a win means a lot of work to get everything else within your control. Look at all the preparation work that goes into getting ready and playing sports at the highest professional level.

Professional athletes spend years acquiring skills and knowledge...but despite that, they're always practicing. They return to pre-season training every year. They review game tapes after each contest. They think, they look at their performance and their team's, and they adjust their plan. Then they train some more.

Unfortunately for many salespeople, you get some basic training on your company's products and services, on company operations, on the employee handbook, corporate policy and procedures and maybe some sales training. Then what?

Yet, companies want or expect superior results from their sales team—but how many take care of the essential training and practice to build and maintain a world class selling organization?

We applaud you for investing in yourself and/or your team, and in your pursuit of becoming an Ultimate Seller.

We are confident that if you will apply the tenets of Ultimate Selling, you will have successful outcomes the great majority of time with your sales opportunities. Of course, in Ultimate Selling, we define successful outcomes as opportunities we pursue that connect, as well as choices made to not pursue an already lost opportunity.

I'm confident if you work hard and apply the tenets of Ultimate Selling and live by its creed, you will most definitely see significant improvement in your sales results; you will know where things stand most, if not all of the time, and you will do so with much less stress, as all the pressure will most definitely be on your prospective customer, not you.

Okay, team. It's time for you to take the field. Go win the game. Go win all of them.

Be awesome. There's nothing that says you can't...and everything that says you can.

Appendix **1**

Personal Improvement Checklist

Congratulations! You've worked hard to arrive here to check your progress and how you rate now.

You progressed from awareness, to knowledge, to application of the knowledge, to ownership of these skills. It's a journey, so refer back frequently for reinforcement to stay at the top of your game.

Let's reflect on your progress to date. You've gained appreciation for a whole new way of approaching sales. You took this quiz upon beginning the book, but you should expect to score differently on it once you start using the Ultimate Selling approach. Please check in regularly and score yourself again—monthly, quarterly, semi-annually, or at whatever frequency will support your objectives.

Know also that we are here for you to help you gain experience and ownership of Ultimate Selling. Don't hesitate to reach out. We'll be honored to help you overcome obstacles. We invite you to become a member of the "Roundtable of Ultimate Sellers", please visit us online for membership details at: **www.ultimatesellingsolutions.com**

Take a time to choose how often you'll retake your quiz and calendar it now:

Monthly   Quarterly   Semi-Annually   Annually

Last Checkup Date: ___________

Where have you made progress?

What do you need to work on next?

Establish your improvement goals and how you will accomplish them:

Refer back to here on your next self-assessment, note how you improved and set some new goals.

Statements	True	I Wish
I only pursue prospects who have high probability for a successful sale.		
My sales pipeline consists of highly qualified, high-return prospects.		
I always meet or exceed my forecasted sales revenue.		
My accuracy is 100% for 30-day sale forecast.		
My accuracy is 85 - 90% for 60-day sale forecast		
I only submit proposals/value propositions to Qualified Prospects…not suspects or unqualified prospects		
I receive acheive a high return of investment for time, efforts and resources invested in prospects.		
I'm never used by prospects to simply gain knowledge which they use to buy elsewhere		
My cost for new business development is decreasing and my profit margins are increasing.		
It's rare that I lose control of the sales process or lose a sale.		
I enjoy the sales profession: It's challenging and fun.		

Appendix 2

Creed of the Ultimate Selling Professional

How will I know I'm a true Ultimate Seller? Simple! When you can honestly answer 'YES' to all the statements below, you are an Ultimate Seller. Congratulations for your dedication to excellence in your career!

- I truly will be a trusted advisor.
- I approach all sales with integrity.
- I approach all individuals with respect
- I pursue long-term relationships.
- I pursue and achieve success by serving others.
- I listen and understand requirements from my *prospect's* perspective.
- I conduct negotiations only with a win-win outcome in mind.
- I am a subject matter expert regarding my company's offerings.
- I offer true market intelligence about the industry I represent.
- I do not focus on the sale; I focus on how my products and services fulfill the needs and wants of prospects.
- I accept personal accountability for all results.
- I achieve a high return for time and effort invested in prospects.
- I'm committed to continuous personal improvement and lifelong learning.

Appendix 3

Business Changes —So You Should Always Be Improving Your Selling Process

When was the last time you looked without prejudice at your sales processes?

The business marketplace changes like everything else in life. Things which once were cutting edge become outdated and ineffective. Selling processes—the set of activities that move people from leads to suspects to prospects to customers, futures or disqualified—need to change over time like everything else.

When was the last time you reviewed your sales efforts with the intention of increasing performance? Do you know the benefits a of specific action that you are doing as you sell—or the negative consequences? Have you been able to determine what you should be doing...but aren't? Do you know what you should *stop* doing?

For many companies, a review of their sales process just doesn't happen until the circumstances are severe. This is because companies see quality engineering as something only related to production and delivery. They don't see selling as something related to process review.

That's unfortunate, since sales are truly the only profit generator a business has. Even if you cut production cost, you have to have the sales! Unwillingness to invest in improved processes is sure to drive failure over time.

It wasn't until late in the evolution of the sales methodologies that it has become more evident that quality control is just as important to apply to sales as it is to any other process. Managers and organizations who are forward thinking realize that excellence in sales processes can be a differentiator in new business development.

So, why has selling been excluded from applying review and re-engineering to their processes? Is it because they see sales only as pushing and personality—when in fact it's a combination of art and science?

And why is understanding this stuff so critical to you?

Because Ultimate Selling focuses on intentional, careful control of the selling process... and before you can completely control the selling process, you need to know what a good selling process should be...and what it shouldn't be.

How Does an Ultimate Selling Process Review Work?

A review looks at your selling process in detail, and then eliminates ineffective activities, adds necessary ones, and updates useful processes.

As we said earlier, things change, and like anything else, your selling process needs to stay current to be at least effective, and at most, a differentiator.

The Ultimate Selling Process Review is a simple tool to help ensure that your efforts will keep you effective at managing highly qualified prospects. It will help you to more and more quickly weed out those which are not going to result in a sale, and prepare for improved interactions with ones which will. While qualifying is all about the prospect, the sales process review is all about sales outcomes.

What are the Criteria?

It's very simple: any activity that is in your sales process must advance the sale's likelihood (from the perspective of the prospect) WHILE AT THE SAME TIME it positions your company as the provider of choice.

If the activity doesn't serve that purpose: stop doing it NOW!

It's not about what feels good to you or keeping an activity because you've always done it that way. Ask yourself instead if it feels healthy, makes sense, and adds value to the process *as viewed by your prospective customer.* Again, if not: what function does it serve in your selling process? If none, stop using it!

Ultimate Selling requires a no-nonsense assessment of every activity of your selling process. It also requires that the objective, of each activity is very clear, and in fact ARE achievable.

You need to validate every single thing you do in the pursuit of new business. The selling process review does not create extra work. Rather, it limits efforts that are inappropriate, ill advised or poorly invested. Doing too much of the latter can actually put you out of business!

Conducting Your Selling Process Review

Here's some good news: you already did much of the work when you created your 'draft' selling process document. Now, you should proceed as follows:

Step 1—Make a copy of your recently created selling process, to be marked up as you review each item by asking the criteria questions described above. The goal is to represent only essential items in the selling process, meaning those that increase your chances for a winning outcome.

Step 2—As you answer the questions, determine if each activity is essential and mark with and E. If an activity is to be removed, mark it with and R. Those about which you are still undecided, mark with a U.

Step 3—Engage others in the process—a peer, a sales manager, departments that support your efforts, customer service or even a customer, to give you their opinions. Does giving potential customers a giveaway item (mouse pad, Frisbee, ice scraper, bottle opener etc.) with your company name on it make sense? Does doing onsite demos seem to close sales? Go through your list again and re-mark items E for Essential or R for removal.

Step 4—Remove all R's. Look at your selling process again.

I would like you to refer to the illustration below which represents how the efforts of many organizations are focused. While the arrows are going generally in a similar direction, a great deal of effort is being wasted by processes butting up against one another.

Our goal is that your organization has highly-focused processes that integrate well. Please take a look at the ideal Ultimate Selling entity represented below.

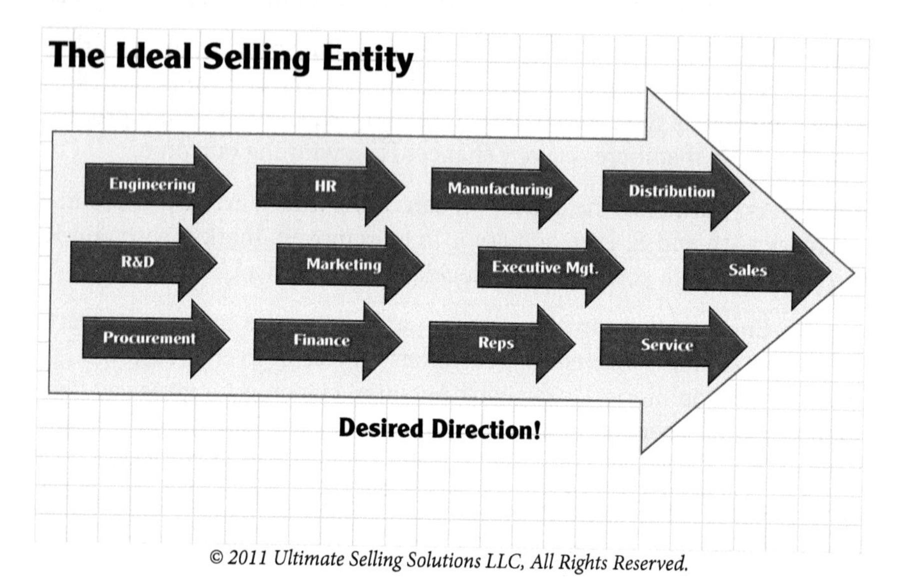

Whenever possible, we invite you to take the initiative with your organization to address aligning processes so that related efforts work together more effectively. This should be a top-down, C-level supported initiative.

Finally, determine how often you should review your selling process, as a stand-alone and for how it integrates with your organization's other processes. Schedule it as a recurring event in your task list. We recommend that you do this annually, as part of a year-end review in preparation for supporting next year's goals.

When would it make sense to conduct a sales process review more frequently than annually? Examples would be:

- if you have a very rapidly-evolving service field or product
- if your company experiences high turnover of sales staff
- if sales are dropping faster than the local economy
- anytime you find that your results and outcomes are short of plan

Once your selling process is updated, keep it handy where you can refer to it often, so that you can consistently use it in creating intentional Interaction Planning with your prospects. Good selling!

Appendix **4**

TENETS OF ULTIMATE SELLING

- Maximum control of the selling process by the salesperson
- Submission of a value proposition/proposal ONLY to fully qualified prospects…not suspects or unqualified prospects
- High return on your investment of time, resources and money in a prospective client or customer
- A formalized selling process for maximizing outcomes, measurement and continual improvement.
- Incorporates keys for success using 'rules of engagements"
- Incorporates thresholds for continuation of the buying/selling process: the seller must continue to meet the needs of the prospective buyer and the prospective buyer must continue to meet the needs and criteria of the seller—for the selling process to continue.
- Focuses on the highest probability for a successful outcome
- Based on the buyer and seller continually earning and enhancing one another's trust and respect.